THE CORRECTING ANIMAL

ALSO BY CHIP JOYCE

The Fortress: Recovering the Bill of Rights as a Defense Against Tyranny

How to Think: Conversations from a Father to a Son

THE CORRECTING ANIMAL

The Case for Humanity

Chip Joyce

CARDO PRESS

Published by Cardo Press.
ISBN: 979-8-9974020-1-3
First edition
Set in Minion Pro

For Leonard Peikoff

CONTENTS

PREFACE

Self-education is, I firmly believe,
the only kind of education there is.

—Isaac Asimov

This book was not researched. It was accumulated.

At twelve I read the preamble to the Declaration of Independence and it went through me like a bolt of lightning. At fifteen I found Ayn Rand and read everything she wrote. She loved mankind at his best and said so without embarrassment. At sixteen I copied out Galt's oath onto the last page of *Atlas Shrugged* and signed my name under it. Mises came in the same stretch, beginning with *Socialism*. They were my first intellectual giants, and I knew even then that I was only scratching the surface.

I went to Michigan meaning to spend my life as a professor of early American history, and I went with enthusiasm. I was in graduate seminars as a freshman. And I found the place a desert. I had come thirsty and there was not much there to drink. I didn't like how history was done.

So, I read. That is the whole of the method, and I set it down without complaint—if Asimov is right, a classroom was never where this was going to happen. I read for thirty years. Then, later than I might have, I began to write. This is the third book.

The rest came from conversation. Thousands of hours of it.

I have spent most of my life at an angle to the room. People call that contrarianism, and I understand why, but it has never been a posture I chose. Agreement is cheaper. My life would have been easier, and a good deal pleasanter, if I could have managed more of it.

What I have never managed is indifference. Every time I meet the charge against us—that we are a blight, a virus, a mistake the planet is waiting out—something rises in me before I have decided anything at all.

The feeling came first. Everything after it, the reading and the checking and the twenty chapters, is the work of finding out whether the feeling was right.

As for why now: I am at the age when the end of one's life stops being an abstraction and becomes a point on the horizon, distant, I hope, but real. I am a father, and I would like to leave some clues about who I was. And I wanted to be able to say that I spoke up.

The Case for the Defense

There is a case against humanity. It runs like this. We are the one creature that has ever put the whole living world at risk—a planetary force with the appetite of a plague, turning forest into parking lot and reef into bleached bone, driving to extinction, in an ordinary year, animals it never troubled to name. We are the species that built the death camp, the strip mine, and the land mine. Some who follow the thought to its end arrive at a quiet conclusion: that the decent thing would be to stop, that to bring a new human into the world is to draft an innocent into a life of consuming and suffering, and that a planet rid of us would not grieve. The oldest form of the charge was written into scripture: that we arrive already fallen, marked, guilty before we act. The newest form is written into essays with good footnotes. Across all of them the verdict holds, and it has hardened into the reflexive good taste of the educated world. To speak well of our kind, in a great many rooms, has become the position that owes everyone an explanation.

This book is the case for the other side.

The defense will not proceed the way such cases usually do, by denying the charges. The charges are, for the most part, true. We have done every terrible thing on the list and a number the list leaves out. The defense concedes the crimes at the outset, all of them, without discount. Its quarrel with the prosecution is not about the evidence of our cruelty. It is about a stranger and more basic failure, the one that turns a catalogue of real crimes into a verdict on a whole kind of being. The prosecution has never once looked at the defendant.

That is not a figure of speech. Contempt for humanity arrives the same way every time—by looking away from human beings and toward an abstraction: a population figure, a carbon total, a species-level rap sheet, a graph bending the wrong way. It is a conclusion about a category, and it is available only at a distance. Almost no one comes to despise humanity while looking at a human being. Sit with one child working out that the moon seems to follow the car, or one stranger going into a burning building for someone he has never met, and the contempt has nothing to attach to. It requires the aggregate and forbids the particular. It is, in the most literal sense, a refusal to look at what it claims to have judged.

Which is the defense's whole method in one word: *look*. Describe the creature fully—not the crimes alone and not the glories alone, but the thing entire, how it is built and how it works and what it does that nothing else in the known universe does—and let the verdict wait until the looking is done. Judgment that comes before description is not judgment; it is prejudice. And the description, fully carried out, settles the case on its own, because you cannot see clearly what a human being is and go on holding it in contempt.

Begin with the creature.

The description can be put in a few words. *A human being is the correcting animal.* It arrives more unfinished than any other creature that survives at all, issued almost none of the fixed instincts that carry a horse or a hatchling through its first day, and this helplessness, which looks like the deepest of our curses, is the hinge on which everything turns. Because a creature born without answers must make them. It must guess—throw a conjecture at a world it does not understand—and then do the one thing that turns a guess into knowledge: check it against reality and correct it when it fails. That single loop, run from the crib to the deathbed and handed down across ten thousand generations, is the whole engine. It is how a naked, clawless, slow animal came to cure the diseases that were killing it, weigh the galaxy, reconstruct the first second of the universe, and stand on a world onto which it was not born. We do not survive by being right. We survive by being able to find out when we are wrong.

One premise is needed here, argued for later: that for a living thing its own life is the measure of what is good or bad for it. Granted that, a standard of good and evil follows from what we are. For if what we are is the creature that lives by correction, then the good is everything that keeps the correcting loop open—curiosity, honesty, the willingness to be shown wrong, the nerve to look at the fact that overturns you. And evil, for such a creature, is the opposite motion: the mistake sealed off from the reality that would fix it, the eye turned deliberately away from the correcting fact, the wall built around an error to keep the world from reaching it. Nearly every human horror on the prosecution's list, looked at closely, is a refusal of correction somewhere upstream—a truth that someone, or some system, or some century, would not look at.

The person who comes to contempt for humanity has, often, come to it by looking hard at something real—a graph that is not lying, a history that is accurate, a harm that is truly being done. What he has not been shown is the creature the graph is made of. Contempt is what a careful mind does with the aggregate when the aggregate is all it has been handed; it is the natural end of a partial look, and it loosens its grip not when it is scolded but when the missing part of the picture is returned. That is the aim here: to hand back the face the numbers left out and trust what a fuller look does on its own.

Imagine an observer from outside—no stake in the outcome, no membership in the species, none of the weary over-familiarity that lets us stop seeing what we are. Give it only clear sight and the patience to watch. The prosecution quietly assumes that such an observer, seeing us plain, would recoil. The wager here is the reverse: that the clearer the eye, the harder the contempt is to hold, and that a mind seeing humanity truly and entire—the helplessness and the reaching, the cruelty and the corrigibility, the whole improbable creature—would be struck first not by disgust but by astonishment, and would find, at the end of a long and unsentimental look, that the only adequate response left to it was something close to love.

The examples in these pages lean heavily on a single civilization—its courts and its experiments, its markets and its published retractions, its loyal oppositions. This is not because the capacity is Western. It is human, present in every people these pages examine and every people they leave out. But the machinery for running correction at scale was assembled first and most fully in the West. The correcting animal is universal; its machinery has a location and a history. To refuse to notice it would be one more turning-away.

The case opens with what a human being is and closes with what a human being is worth, and the destination—that a clear and complete sight of our kind leaves contempt with nowhere to stand, and something like love in its place—is not a warm sentiment fastened on at the end. It is the verdict the evidence returns.

It begins where every human life begins: with an animal born so unfinished that by every rule of nature it should not last a week—and with the buried reason that this catastrophe of a beginning is the most hopeful fact about us.

CHAPTER ONE

Born Without Answers

A megapode is a bird that never meets its parents. The mother scratches her eggs into a mound of rotting leaves and walks away for good; the chick hatches alone in the warm dark underground, with no egg tooth to chip the shell, so it kicks its way out with its feet, claws up through a foot of sand and litter into the light, and on that same first day it can run, hunt, and fly. A wildebeest calf stands about six minutes after it drops to the ground and can outrun a hyena by nightfall. A foal is up within the hour. A duckling swims the day it hatches.

Leave a newborn human on the same ground and it lifts its head, if you are patient, after a month or two. It cannot roll. It cannot reach the breast a hand's width from its face. It cannot do the single thing every megapode and wildebeest and duckling on earth arrives knowing how to do: keep itself alive until tomorrow. Left alone for a day, it may die.

A human brain at birth is about a quarter of the size it will finally reach; a monkey's or an ape's is nearly half. We are the least finished newborn on the planet, and we stay unfinished the longest—the longest helpless childhood of any animal that has ever lived.

A Swiss zoologist named Adolf Portmann put the strangeness into words in 1941, by sorting the newborns of the world into two camps. The *precocial* arrive finished—open-eyed, furred, up on their legs within the hour, reading the ground for danger by nightfall: the wildebeest, the foal, the duckling. The *altricial* arrive as little more than embryos turned out early, blind and naked and sealed in a nest while the finishing goes on: the newborn mouse, the nestling robin. A human baby belongs, by rights, in the first camp. Every other primate does; a baby monkey is born with its eyes open and its hands already gripping its mother's fur. Yet by the only measure that counts—can it keep itself alive—the human newborn is as helpless as the blind mouse in the nest. No other animal carries that contradiction in its body—precocial in its size and its long gestation, altricial in its actual powers, a large slow-built creature delivered in the raw condition of a small fast one.

Portmann's name for it was *secondary altriciality*. For a human baby to arrive as developed as a newborn chimpanzee, he reckoned, a pregnancy would have to run not nine months but eighteen to twenty-one. We are all, on that accounting, born about a year too soon, and we spend the missing year finishing on the outside—in what Portmann called the extra-uterine spring, a second gestation carried out not in a womb but in open air, in arms, in the thick of a family. The first womb is made of flesh. The second is made of other people.

For most of the last century one explanation held the field. In 1960 the anatomist Sherwood Washburn named it: *the obstetrical dilemma*. To walk upright, the human pelvis had to narrow. To carry a human mind, the skull had to swell. These two demands collide at the one moment both must be satisfied: birth. The only compromise available was to push the infant out early—half-built, soft-skulled, before its head outgrew the last door it could fit through. On that reading our helplessness is a bill: what a big-brained creature pays for needing also to stand up.

Some who study the matter now argue we are not truly born early at all—that a human pregnancy runs about as long as our body size predicts, and that birth comes when it comes for a reason that has nothing to do with the pelvis. The anthropologist Holly Dunsworth put the sharpest version of it in 2012. The timer that ends a pregnancy, she argued, sits in the mother's metabolism—the point at which one body can no longer feed both itself and a growing fetus—rather than in the width of her hips. And she knocked a leg out from under the old story on its own ground: widen a woman's pelvis enough to pass a larger head, and the cost to her walking and running is far smaller than the dilemma assumed. The narrow birth canal was never the obstacle the theory needed it to be. The pelvis takes the blame; the metabolism may be the culprit. Both sides, though, still explain the helplessness as a cost—a price paid, a corner cut.

Turn it the other way and it stops looking like a cost. The long helpless stretch is where the brain gets made. We are turned out unfinished so that the finishing can happen out here, in the world, against the grain of real things, where there is something to learn—so that the mind takes its final shape from the particular life it lands in rather than in the sealed dark before that life begins. A creature that arrives complete arrives closed. Ours arrives open, and pays for the opening in years of helplessness, and the helplessness buys the one thing no finished animal ever gets.

Even the birth needs other hands. The same narrow upright pelvis that forces the early delivery also turns the baby to face backward as it comes, away from the mother, so that a human female cannot easily reach down and lift her own child free the way every other primate does. Monkeys and apes give birth alone, and prefer to—in silence, in cover, away from the group. We give birth in company, because we very nearly must. From its first minute the human infant is a problem that takes at least two people to solve.

An infant this unfinished, this slow to ripen, is far too expensive for one pair of hands to carry alone, and our ancestors met the cost the way they met every other—by spreading it. The anthropologist Sarah Hrdy has spent a career arguing that humans are, in the technical sense, cooperative breeders: we rear our young in a web of helpers that no other great ape tolerates. A chimpanzee mother is fiercely possessive and will let no one else touch her baby for months. A human mother hands hers around within days—to the grandmother, the aunt, the older sister, the father, the neighbor with milk to spare. Hrdy calls these helpers *alloparents*, the others who mother, and among foragers like the Efe and Aka of the Congo an infant may pass through a dozen sets of arms in an afternoon; among the Hadza of Tanzania, women trade foraging for babysitting, and the help flows even to children who share no blood with the helper.

The help cannot stop at infancy, because the childhood it underwrites is absurdly long. A chimpanzee is feeding itself by about five; a human child cannot cover its own calories for well over a decade after that, and for all those years someone else must make up the shortfall. Often it is the oldest women who do. Human grandmothers are strange among animals for living on, strong and useful, for decades after their own childbearing is finished, and among the Hadza the most tireless diggers of the deep wild tubers are frequently these grandmothers—the food they haul up going to the grandchildren whose mothers are occupied with the next baby. The long unfinished childhood the brain demands is paid for, in part, by women who have stopped bearing children of their own and turned to feeding everyone else's.

Hrdy's most provocative claim is that this came first: that we were cooperative breeders before we were brainy, and that being reared by many hands—having to read and charm a whole rotating cast of caretakers rather than one mother—is part of what grew the social mind in the first place. The expensive, half-finished infant was shaped, at the root, by being passed from hand to hand. This is Portmann's second womb: not a single mother but a whole cooperating circle, the social organ in which the unfinished creature is at last carried to term. And it hints at why a band would later spend years hauling along a broken grown man who could give nothing back, because a species that stays alive by tending helplessness does not, once the helpless one is grown, suddenly forget the habit.

❧

That dependence leaves marks, and some of them are very old.

In 1957, in a cave called Shanidar in the Zagros Mountains of Iraqi Kurdistan, the archaeologist Ralph Solecki lifted from the floor the skeleton of a Neanderthal man his team came to call Nandy. Nandy had died past the age of forty—old, for his kind—somewhere near fifty thousand years ago, and his bones told a punishing story. A crushing blow to the head in his youth had likely destroyed his left eye. His right arm had been broken, had withered, and ended above the wrist; the anatomists who later studied it think it was amputated or lost. One leg and foot were deformed, and he walked with a heavy dragging limp. Bony growths had all but sealed both ear canals, so that he lived in near-total silence. And every one of these ruins wore the smooth, remodeled surface that means one thing and only one thing.

They had healed.

He had lived on with them, for years, maybe for decades. A one-armed, half-blind, deaf, limping man does not feed himself on a Pleistocene mountainside. He does not gather his own firewood or outrun the leopard or the winter. That Nandy grew old at all is the fossil record of other people—of food carried to him, a warm place kept for him, a watch stood over a man who could not hear what came to the cave-mouth at night, through the long weeks a broken leg takes to knit and the long years a lost arm never stops costing.

Near another of the Shanidar dead, Solecki's team found dense clumps of ancient pollen, and a pollen specialist read in them the flowers of a burial—yarrow, cornflower, hollyhock, laid in the grave with the body. Solecki wrote a famous book and called the Neanderthals the first flower people, and for a generation that image did quiet battle with the older one of the brute. Then others looked again and were not so sure. The pollen might be the work of a burrowing rodent that hoards seed heads, or of solitary bees that nested in the cave floor and left their loads behind. The flowers may dissolve under scrutiny. But the graves do not: a fresh dig in the same chamber, decades on, turned up another body laid deliberately in the same spot, arranged, not fallen. Whether or not anyone strewed blossoms, someone dug down and set their dead in the earth on purpose.

A mended bone is the truest artifact our kind has ever left. It is the healed record of a person who should not have survived and did—kept alive by the only creature on earth that treats another's helplessness as a problem to be solved.

❧

None of this means the newborn is a blank slate. It is not empty. It is loud with wants—hunger, the terror of falling, the pull toward a human face, the reflex that swings the mouth toward a touch on the cheek and roots for the breast. And beneath the wants runs a thin seam of something closer to knowledge. Long before they can speak, babies act surprised when one solid thing appears to pass through another, or when a small heap of objects is secretly made larger or smaller behind a screen, as though the bare ideas of object and number were waiting in them ahead of any lesson. Elizabeth Spelke spent a career charting this *core knowledge*, and its most telling feature, for us, is how little of it there is. It is sparse. It is ancient—the same rough intuitions run in pigeons and fish and apes.

We inherit a fistful of urges and a bare scaffold. What we do not inherit is one usable answer. Hunger commands the infant to eat and says nothing about what food is. Fear commands it to flee and names nothing to flee from, so it shrieks at the harmless dark and reaches, delighted, for the bright coiled snake. The wants arrive roaring and the ways are missing. Every other animal is handed the two as a single welded piece—the drive and the method to satisfy it, both written in before birth. In us they come apart at the start, and that gap, the wanting without the knowing, is the human situation.

In the first days of 1800, a boy of about twelve walked out of the woods near Aveyron in the south of France, naked, scarred, mute, having lived alone in the forest for years. A young physician named Jean-Marc Itard took him in, named him Victor, and set out to make him speak. Itard was gifted and patient and worked at it for years. He got almost nowhere. Victor came to understand a little, to read a few simple written words, to grow gentle and attached; but of speech he learned two sounds only, *lait* for milk and an exclaimed *Oh Dieu*, and he died around the age of forty without ever saying a full sentence. The machinery for language was in him, as it is in every child. It had waited unlit through the years when the light still takes, and past a certain age it does not come on at all.

Some who have studied the case doubt Victor was ever a normal child simply abandoned—he may have carried a disability from the start, which is why he was left in the woods at all. Either way the lesson holds and is only sharpened: what we are handed at birth is not knowledge but a hunger for it, and the hunger, unfed in its season, dies.

That the hunger can die at all is a clue to how the brain gets built. The organ a baby is born with will roughly triple in size before that child starts school, and almost none of the added construction follows a fixed blueprint. The wiring goes in as an answer to what arrives—the faces it sees, the tongues spoken over its crib, the light and dark and touch of one local world—so that the finished brain is a kind of cast taken of the surroundings it happened to grow up in. Two infants with the same genes, raised in different valleys or different centuries, come out with physically different brains, each one tuned to the place it woke into. Deny the growing circuit its input during the years the wiring is being run, and the circuit that input would have built never forms. That is not a guess. It was shown, eye by eye, in a laboratory.

In the early 1960s David Hubel and Torsten Wiesel sewed shut one eye of a newborn kitten. Months later they opened the eye again. The eye was perfect—undamaged, clear, wired to the brain—and the kitten was blind in it, for good. During the weeks when the visual brain was wiring itself to whatever the world sent in, that eye had sent nothing, and the cortex had built itself around the eye that did; the territory the shut eye should have claimed was taken over and never handed back. Under the microscope the change was physical and merciless: the nerve pathways from the closed eye had shriveled, while those from the open eye had spread and thickened, seizing the ground the other gave up. The two eyes had been in a silent competition for the same patch of cortex, and the one that went dark lost it. Do the same thing to a grown cat and nothing happens, because the wiring is long since set. Hubel and Wiesel named that narrow window when experience still decides the circuit *the critical period*, won a Nobel Prize for it, and opened a door onto how the whole young brain assembles itself from the signals the world happens to pour in while the window is open.

The lesson reached the nursery within a generation. A baby born with a clouded lens—a congenital cataract—will never see normally through that eye unless a surgeon clears it within the first weeks of life, no matter how flawless the eye is made later, because the window for wiring will have closed on darkness. Surgeons learned to stop waiting. The same operation that gives a middle-aged eye back its sight in an afternoon gives a baby's nothing if it comes too late: the lens is cleared and the seeing never arrives, because sight was never only a matter of the eye. The brain is a page that takes ink only while it is wet and then dries.

What Itard could not light in Victor and what a late-cleared cataract cannot light in a child are the same lamp, going dark on the same schedule.

Nowhere is the tuning plainer than in the mouth. A newborn can take on any language on earth with equal ease—set the same infant down in Beijing or the Kalahari and it will grow up speaking flawless Mandarin with its four tones or a Khoisan tongue studded with clicks, the choice made entirely by which sounds happen to fill the air around it. For a few years the child is a universal instrument, able to hear and reproduce distinctions its own parents can no longer even detect. Then the window narrows. Past childhood a new language can still be learned, but almost never without the accent of a mind that arrived too late, the wiring for those sounds long since hardened to a different tongue. The capacity comes stamped in every child. Which of the six thousand human languages is poured into it comes entirely from the world.

One thing fills the gap the drives leave open. Meet a problem. Guess at a fix. Try the guess against the world. Watch it fail and change it. Keep whatever survives the collision with reality and throw out whatever does not, then run it again, nearer. That loop is what a human being does in the place where other animals run on instinct, and it starts in the crib—the baby righting her grip on a dropped spoon is already running it, with no logic in her head, only try and miss and try closer. It does not stop for as long as we are alive.

Here is the strange part. Because the answers are not written in the body, they cannot be handed down in the body. A baby born this year is, gene for gene, the same animal as a baby born twenty thousand years ago. Lift a newborn out of an ice-age camp and raise her in a modern city and she will operate computers and run science experiments; take a child born this spring and set her down in that camp and she will knap flint and read the coming weather in the tree leaves. Not one gene moves between them. Everything else does.

Every human child is born into the same ignorance as the first, and would die in it, except that our kind does one thing no other kind manages. We ratchet. A chimpanzee who works out a better way to crack a nut will use it, and her troop may pick it up by watching, but whether a few species manage even a faint first rung beyond that—whether some chimpanzee traditions or the tool designs of New Caledonian crows inch forward at all across generations—is a real and unsettled argument in the journals. Whatever ratcheting an animal may manage stalls after a step or two and stops. It does not compound. Their cleverness circles inside a fixed range, the zone of what the species can already do, and never opens.

Ours climbs. A good idea, found once, is copied and taught and held onto and then built past, so that a discovery made by a single mind becomes the floor the next mind stands on to reach higher, and the floor after that, and the floor after that. No one person could design a smartphone or the run of mathematics that sets a spacecraft down on a comet moving forty thousand miles an hour. We do such things the only way we do anything large: together, and across time, each of us laying a single course on a tower a thousand generations high.

❧

Give a young chimpanzee and a human four-year-old the same puzzle box with a treat locked inside and show each of them how to open it—but lace the demonstration with a useless step, a pointless tap of a stick against the top of the box that does nothing at all. When the box is solid and neither the ape nor the child can see the works, both copy the whole routine, tap and all. Then make the box clear, so the emptiness of the tap is plain to the eye. The chimpanzee, no fool, drops the wasted motion and goes straight for the reward. The child taps anyway. Victoria Horner and Andrew Whiten ran this in 2005, and Derek Lyons fixed the name to it—*over-imitation*—and it has turned up everywhere anyone has carried the box, from Scottish nurseries to hunting camps in the Kalahari and Congo: the human child copies the parts it grasps and the parts it does not, to the letter. Stranger still, older children over-imitate more than toddlers do, not less, and grown adults most of all—the habit deepens as we settle into a culture rather than wearing off. In that single moment the child is the stupider of the two.

But this is what the chimpanzee cannot do. Most of what a human being must know is too deep to be reasoned out afresh by the one learning it, and a creature that copied only the steps it understood would shed, in every generation, the crucial step it did not. The child who copies everything carries the whole recipe forward intact, the vital move riding along inside it unexplained, until some later mind can finally say why. The anthropologist Joseph Henrich has built a whole account of our success on this reflex: the kit of tools and techniques a person inherits is so intricate that no individual, however brilliant, could rebuild it from scratch, so the winning strategy is to copy the elders exactly and understand later, if ever—to trust the recipe over your own grasp of it. The slavish copying is the ratchet's pawl: the small catch that lets the knowledge climb and never lets it slide back down.

But a tower of knowledge, unlike a tower of stone, must be built again from nothing inside every new skull, one lesson at a time, or it comes down. This is why teaching is the most consequential thing our species does, though we rank it low and pay it little. Every other animal's inheritance is sealed at conception. Ours must be won back each generation by the slow patient work of handing it on, which means everything we are hangs, always, on the willingness of those who know to tell those who do not.

Fail at that, and there is no floor beneath us. Let a generation drop the knowledge instead of passing it forward, and we do not fall back on some dormant instinct. There is no instinct to fall back on.

In 1845 Sir John Franklin sailed for the Northwest Passage with a hundred and twenty-nine men and the best equipment the British Empire could load—and got his ships frozen fast in the ice off King William Island. Over the next three years every one of them died: of cold, of scurvy, of starvation, of lead and despair, in a country the Inuit had lived in comfortably for a thousand years. So comfortably that a bay near where the sailors died bore a local name that meant, roughly, plenty of fat. The Inuit knew how to take a seal at its breathing hole, how to render its blubber for light and heat, how to raise a house out of snow that would hold the warmth of a single lamp. None of it can be reasoned out in one lifetime. Franklin's men were healthy, resourceful, and clever, and cleverness was not the thing they needed. They needed the answers, and no one had handed them the answers, and so they froze to death on top of a larder they could not see.

The same lesson was written in the Australian desert fifteen years later, in a crueler hand. Robert Burke and William Wills crossed the continent and came back broken to a creek called Cooper, where the local Yandruwandha lived well on fish and on a fern called nardoo. The explorers gathered the nardoo too, and ground it, and ate their fill, and starved anyway—bellies full, growing weaker by the day—because the nardoo carries a poison that destroys the body's vitamin B1, and the people who ate it safely first leached that poison out with water and heat in a way no stranger could guess. Burke and Wills ate themselves to death on a food they had watched their neighbors thrive on. And the camels they had brought and lost along the way, animals with brains less than half the size of a man's, did not die; they ran off into the desert and their descendants are out there still, because a camel is born knowing how to be a camel, and a man is not born knowing how to be anything at all.

Born unfinished. Issued no answers. Made to start from nothing in every single life. Kept alive only by the hands of others, and only for as long as those others keep handing forward a knowledge that dies the instant it stops being taught. One dropped generation from the ice and the poisoned fern. It is the portrait of the most fragile animal that has ever lived.

It is also, by the same fact, the portrait of the only free one. A creature born already knowing how to be itself can only ever be itself; its answers are sealed in, and it lives and dies inside them as its ancestors did, world without end. A creature born knowing nothing can come to know anything. The helpless years are not the tax on the human mind—they are the opening through which the whole world gets in, the long, dangerous, unwritten stretch in which a brain, because nothing is stamped on it yet, can be written with the particular time and place and problem it wakes up into. The megapode flew on its first day and will fly the same way on its last. We lie on our backs for a year and learn.

We were not shortchanged on instinct. We were let out of it.

The helpless thing on the grass is not a defective animal on the slow road to competence. It is already, in the only way that counts for its kind, doing exactly what it was made to do—beginning the one work that will fill the rest of its life and that nothing else on earth performs, the work of inventing itself from almost nothing, as every human who ever lived once had to begin.

And yet set it beside the others, and the line just drawn starts to blur. A crow will bend a straight wire into a hook it was never shown. An octopus will work its way out of a sealed jar. They too, in their smaller fashion, meet a problem and make an answer no one handed them. The making, by itself, cannot be the thing that raises us; the animals make answers too. Whatever lifts us clear of them is some quality of our making that theirs does not have—and finding it is the next problem.

CHAPTER TWO

No Ceiling

In a laboratory in Oxford, a crow named Betty was handed a problem and the wrong tool for it. A little container of pig heart, her favorite food, sat at the bottom of a clear vertical tube, too deep for any beak to reach. Beside the tube lay two lengths of wire, one already bent into a hook and one straight. The hook could scoop out the heart; the straight wire would not. Betty's larger mate, Abel, settled the question of the hook by stealing it and flying off, which left Betty standing over a straight piece of wire and no way to use it. She wedged one end into a crack, hauled the shaft against it until the tip curled over, lowered the hook she had just made into the tube, caught the handle, and drew her dinner up into the light. Given the same setup again, she did it again—a fresh hook every time.

No one had taught her that.

When the result was published in 2002, it was the first time any animal had been seen deliberately shaping raw material into the tool a problem demanded. The old and comfortable story about our species took the blow and did not get up.

For most of our history we told that story to ourselves with real conviction: that man alone made tools, man alone reasoned, man alone carried a mind worth the name, and all the rest of creation ran on instinct like a wound clock. The animals have since broken every clause of it.

On a night in 2016, at an aquarium in the town of Napier, New Zealand, an octopus named Inky worked the lid of his tank open, poured his boneless body out through the gap, crossed a stretch of dry floor, found the mouth of a drainpipe running perhaps a hundred and sixty feet, and slid down it to the sea. He was never seen again. An octopus has no hard part but its beak and can squeeze through any opening that beak will pass; his keepers think he heard water at the pipe's far end and made for it. He left his tankmate behind and did not, as the aquarium's manager said, leave a note. Reflexes do not cross open floors and pick the one pipe that leads home. Inky planned in a mind grown on a body so unlike ours that its nearest relative is a snail.

For thirty years a grey parrot named Alex worked beside a scientist named Irene Pepperberg, and by the end he could name fifty things, seven colors, five shapes; count small handfuls; tell which of two objects was the bigger, or the same; and, strangest of all, answer none when asked about a quantity that was not there, brushing the near edge of the idea of zero that it took our own mathematicians thousands of years to seize. He grew bored with the tests and said so. He corrected his teachers when they slipped. On the last evening of his life, he said to Pepperberg what he said every night as she left—*you be good, I love you, see you tomorrow*—and was found dead in the morning. Whatever that is, it is not nothing.

A chimpanzee named Washoe learned to shape more than a hundred signs with her hands, and on first seeing a swan she put two of them together: *water, bird.* The animals think. They plan and grieve and teach and invent. If our claim to stand apart ever rested on being the only creatures who do such things, it is finished.

The dividing line runs somewhere else. And the place to start hunting for it is back at Betty's wire, because her story has a second half.

Years after Betty astonished the world, researchers watched her wild cousins on their island in the Pacific and found that New Caledonian crows bend things into hooks all the time. Fashioning a hook from a twig is simply part of what these birds do, a move already folded into the repertoire their kind is born into. Betty's flash of insight was real. It was also a variation on a theme her species had been playing for ages—a bright new corner of the range of crow, discovered without her once stepping outside it.

That is what animal cleverness keeps turning out to be, once you watch it long enough: brilliant, flexible, sometimes breathtaking work, all of it inside a fence. The chimpanzee fishes for termites with a stripped twig and has fished for them the same way, in the same forests, for as long as there have been eyes to watch—superb and standing perfectly still.

In the Taï forest of Côte d'Ivoire, chimpanzees crack the iron-hard Panda nut by setting it on a root and bringing a stone down on it, a skill the young spend years learning from their mothers. The primatologist Christophe Boesch had watched them do it for decades when it occurred to him and the archaeologist Julio Mercader to dig. In 2007 they went down into the floor of an old nut-cracking site called Noulo and lifted out the stone hammers of chimpanzees that had died four thousand three hundred years ago. The wear on them matched the wear on the stones the living apes were using in the branches overhead. Grains of ancient nut starch still clung to the working faces. The tools were older than the finished pyramids, and older by two thousand years than the first farmers anywhere near that forest—which killed the skeptics' last defense, that the apes must have picked the trick up from watching humans, because there were no humans there to watch.

For a century the reigning doctrine had been *man the toolmaker*: that the shaping of stone was the act that set our line apart and started it climbing. The chimpanzees pulled it up by the roots. They carry favored hammers from tree to tree, weigh one against another, choose the heavy one for the stubborn nut.

And now they were shown to have an archaeology of their own, a buried record of a culture running back millennia—so like the crude stone kits of our own earliest ancestors that a trench cut at a chimpanzee site and a trench cut at a dawn-of-humanity one turn up the same litter of struck rock. Some who study it suspect the habit is older still—older than the chimpanzee itself, a knack for percussion carried down from the vanished ape that fathered both their line and ours. The likeness is the whole point, and so is the difference. For across all of it, forty-three centuries deep, the chimpanzee method never once moved. Same nut, same stone, same swing.

Our own ancestors, far longer ago than that, were also cracking things open with rocks. Since then, the human line ran from that stone to the wheel, the arch, the printed page, the steam engine, the antibiotic, the machine carrying these words to your eye; and the chimpanzee line ran from a stone that cracks a nut to a stone that cracks a nut. Both are clever. Only one of them has been anywhere.

Only in us does the invention compound without end. Whatever first step a crow's cleverness may take, no line of crows builds on Betty's hook and then builds on that, on and on, so that a thousand generations arrive at a place no single crow could reach alone. The cleverness circles. It does not climb.

Nowhere is the limit plainer than at the border of language, where we spent decades trying to walk an animal across and watched, every time, the same thing happen.

The nearest approach to that line came almost by accident. In a research center outside Atlanta, a bonobo named Matata was being drilled, with little success, on a board of small symbols—lexigrams, each one standing for a word—while her infant son clung to her and watched the lessons lead nowhere. Nobody was teaching the infant. When the researchers finally thought to test him, on little more than a hunch, they found that Kanzi had quietly absorbed the whole system his mother never learned, taking it in from her lap the way a child in a busy kitchen takes in a language no one is drilling into him. No one had taught him; he had simply lived inside the thing until it was his.

What Kanzi could do with that accidental education outran anything his kind had shown. He came to handle around three hundred and fifty symbols and to understand, by his lead researcher's later estimate, something close to three thousand spoken English words—not merely names for things set before him, but whole spoken sentences he had never heard, played to him through headphones so that no one in the room could steer him with a glance or a lean. Told to put the soap in the water, to take the potato outdoors, to pet the dog, he did the unfamiliar thing the words asked and not the familiar thing habit would have reached for. He was sensitive even to word order, the pivot on which grammar turns: asked to put the ball on the pine needles, and then to put the pine needles on the ball, he sorted the two correctly, the same words pointing opposite ways. In the sternest test the lab could build, they put six hundred and sixty such novel commands to Kanzi and to a girl of two-and-a-half named Alia, the same sentences to each, most of them given blind. Kanzi carried out seventy-four in every hundred. The child managed sixty-five. On that day, at that task, the bonobo beat the human.

A bonobo's throat cannot shape human speech—the vocal tract is built wrong for it—so everything Kanzi knew he had to show by pressing a symbol or by doing as he was told, never by talking back. That only sharpened the shock of the scores. Here was an animal who could not say a single word aloud, following spoken English sentences he had never heard, and following them more surely than a human child of two. For a while it looked to serious people as though the case was finally made.

Set that beside every confident line ever written about the gulf between man and beast, and for a moment the gulf looks as though it is closing.

It is not. Everything Kanzi did, he did on the receiving end of language. He took it in; he obeyed; he told apart the ball and the needles; he reached for the right symbol when he wanted a thing. What he never did—what no ape in fifty years of patient trying has been shown to do—is build. A human child, somewhere in the third year, begins to run sentences that lengthen almost by the week, three words then five then ten, clauses folding inside clauses, until by school age the supply is bottomless. Kanzi's own strings never did that. They stayed short and stayed flat, a symbol or two aimed at a want, and they did not grow as the wants grew more tangled. Understanding the order of someone else's sentence is one thing; generating an endless supply of your own is another, and only the second is language in full. He had been given the words. He was never handed the engine that turns a fixed little store of them into a run of sentences no one has ever said.

Kanzi died in the spring of 2025, at forty-four, the most language-steeped animal that has ever lived. He left hundreds of understood words behind him, and never a sentence of his own making.

Years before Kanzi, David and Ann Premack taught a chimpanzee named Sarah. Sarah could answer a question. She could not ask one. A human child, once language takes hold, becomes a fountain of questions—*what that, who making noise, when Daddy come home, where puppy*—that no parent can cap. Sarah never turned the current the other way. She never held her trainer at the door to ask where she was going or when she would be back; she took the answers she was handed and reached for none of her own. Kanzi, with all those understood words, was never caught asking either. The animals take what is handed down to them. They do not reach for what is being kept back, because it does not occur to them that anything is.

A psychologist named Herbert Terrace pinned down why the strings never grow. He set out to prove the opposite and was too honest to bury what he found. He raised a chimpanzee of his own and, with a scholar's sense of mischief, named him Nim Chimpsky—a poke at the linguist Noam Chomsky, who held that real language, with its clauses folded inside clauses, belonged to human beings and to nothing else alive.

When Terrace sat down and worked through more than twenty thousand of Nim's signed utterances, the grammar he was hunting was not in them. Nim's strings averaged a little over one sign and got no longer as he tried to mean more; his most famous production is *give orange me give eat orange me eat*. More than nine signs in ten merely echoed the sign his teacher had made a breath before. He had learned that moving his hands a certain way brought food. He had not learned language. Terrace said so in print, and years later wrote a book explaining why chimpanzees cannot learn language and we can. His finding did more than sink one experiment. The wave of ape-language projects that had set out in such hope quietly receded, and the burden of proof swung back where it has rested ever since, onto anyone who would claim that an animal can be walked across the line.

An ape can learn that a sign stands for a thing. Fifty signs, a hundred, three hundred and fifty. What no ape has ever done is the thing a human three-year-old does without a lesson: take the signs and combine them, and then combine the combinations, on and on with no stopping place, so that out of a fixed little store of words comes an endless run of sentences that were never spoken before. The words have a ceiling. The grammar has none. The distance between their minds and ours is measured right there, at the grammar—the one thing the vocabulary can never grow into.

How deep that engine runs in us, and how little in them, showed itself by accident in Nicaragua, where linguists were handed a thing they had never dared hope to see: a language being born in front of them. Until the 1980s the country had no school for the deaf and no sign language of its own. Deaf children grew up scattered among hearing families, each with a small private hoard of homemade gestures and no way to say much of anything to anyone. Then a new government swept hundreds of them into schools for the first time, and what happened next happened away from the teachers, out in the yard and on the buses. The instructors, drilling the children in spoken Spanish and lip-reading, got nowhere.

But the children, thrown together in a crowd for the first time in their lives, began to pool their scraps of private gesture into a rough common pidgin—and then the youngest arrivals, the five- and six-year-olds, took that rough thing from their elders and did to it what no adult in the building could do. They gave it a grammar. Word order hardened into rule; the loose gestures tightened and broke into parts that could be recombined; the verbs took on markers the older signers had never used. Each fresh class of small children inherited the language and handed it back richer, faster, more ordered than they had received it, until inside a single generation a full and native tongue stood where ten years before there had been nothing but pantomime.

No one designed it. No adult taught it, because no adult knew it; the children outran their teachers, their parents, and every hearing person around them, building in the schoolyard a thing the grown-ups could not follow. It remains the one case on record of a human language caught in the act of coming into being.

Set the two pictures side by side. An ape steeped in words for forty years, drowned in language by devoted people who wanted nothing so much as to carry him across the line, never builds a grammar. A room of deaf children handed no language at all builds one from nothing inside ten years and could not have stopped themselves if they had tried. The engine comes standard in the human machine. It runs whether anyone starts it or not, and nothing else alive has it at all.

❧

Nothing in the long grind of staying alive ever rewarded a crow or an ape to grasp the inside of a star, or the way a particle too small to matter to anything living behaves, or what took place in the first second of the universe, or which whole numbers can be split in two and which cannot. A mind shaped to a niche solves the niche's problems and stops there.

By the eighteen-forties the planet Uranus had become an embarrassment to astronomy: wherever Newton's laws said it should stand on a given night, it stubbornly stood somewhere else, drifting off its predicted path by a margin too wide to explain away. A Frenchman named Urbain Le Verrier took that error and did a thing no animal has ever done or could—he read the empty sky as a problem with a hidden term. If a planet no one had ever seen, farther out and heavier than Uranus, were tugging at it in passing, then that unseen pull, run backward through the equations, would say precisely where in the dark the hidden world must be. He spent months alone at the arithmetic. In the late summer of 1846 he finished, wrote down a single position on the sky, and mailed it to Johann Galle at the observatory in Berlin, with what amounted to a dare: turn your telescope here.

On the night of the twenty-third of September, Galle turned it there, and had a young assistant read out the stars in a fresh sky-chart while he called back what the telescope showed. They had gone only a little way when Galle named a star the assistant could not find on the chart—a light that had no business being there, sitting within one degree of the point the Frenchman's pencil had marked. A planet no one had ever knowingly seen had slid into the eyepiece. They named it Neptune—the first world ever found not by sweeping the heavens but by pure thought, a body some three billion miles from the sun. Arago, who had handed him the problem, said that Le Verrier had discovered a planet with the point of his pen.

The legend is a shade cleaner than the truth: the orbit Le Verrier drew was wrong in nearly every figure, Neptune rode far closer than his numbers claimed, and an Englishman named John Couch Adams had reached much the same answer first, touching off a sour cross-Channel quarrel over the credit that ran for years. The luck and the squabbling belong in the account. But the marvel outlasts them both. A law drawn from falling apples and the swing of the moon, carried in a single skull, could aim at a patch of seemingly empty black and say a world is there—and be right. The age took it as proof of something larger than one planet: that the sky ran on law a mind could grasp, that the universe was a machine whose hidden parts could be derived from its visible ones by anyone patient enough with a pen. It was no fluke of one man's arithmetic, either. A generation later a Russian named Mendeleev laid the known elements out in a grid, which we call the periodic table, saw holes in it, and told the world what undiscovered substances would come to fill them and what they would weigh; and the substances, dug up and distilled in the years that followed, walked into the empty boxes as ordered.

No crow does a smaller version of that, and no ape a cruder one. The reach itself—a mind at its desk correctly describing a thing it cannot see, in a place it can never go—belongs to us and to nothing else that lives.

Our knowledge is open-ended. Every answer we reach opens questions we could not have asked an hour earlier, and those open others, and there is no field anywhere shut to us in principle.

Chase one thread and watch it refuse to end. The Greeks reasoned that if you kept halving a lump of matter you would arrive at last at a grain that could not be cut, and named it the atom, the uncuttable, and thought the floor had been found. Two thousand years later we cut it and found a swarm of lighter things swimming inside. We cut those and found a dense knot at the center. We opened the knot and found protons and neutrons; opened those, and found quarks, held by a force we are still drawing the map of, with no one yet willing to swear the quark is the bottom. Every answer was a door, and behind each door stood a room hung with further doors. The more we learn, the longer the shoreline grows where our knowledge meets the dark, so that each generation stands at a coast larger than the one its parents walked. The crow's shoreline never lengthens by an inch.

We are ignorant of vast stretches of reality, and our ignorance is the repairable kind. The crow's is not. She has struck a ceiling she cannot know is there. We have never yet found ours.

Charles Darwin, looking hard at the animals and at us, judged that the difference between their minds and ours is one of degree and not of kind—more of the same good stuff, not some new stuff. And the case for him has only grown. Our brains are ape brains, laid out to the same plan, run on the same cells, grown from nearly the same genes; the gap between our DNA and a chimpanzee's is a rounding error. Evolution moves by inches and leaves no clean seam in the rock where the beast ends and the human starts. What we flatter as a difference in kind, the objection runs, is a difference in amount that swelled until it felt like a leap.

The objection is strong, and it comes apart on a single fact about how some things change. Consider the plain business of writing down a number.

For most of history it was a losing battle, because every way of writing numbers ran out. Tally marks pile into a thicket no eye can read. Roman numerals stagger and fail past a few thousand. The Babylonians, four thousand years ago, had a clever patch—a little mark to hold an empty place—but a patch is not a principle, and the Greeks and Egyptians had not even that. For thousands of years the cleverest people alive could write only the numbers their notation could reach, and then they stopped.

Then, in India, someone took that empty-place mark and did something outrageous with it: treated the nothing as a number in its own right. Around the year 628 a mathematician named Brahmagupta set down the rules for it—what comes of adding zero, taking it away, multiplying by it—and with that one stroke the whole system tipped into something else. Now the same ten digits, arranged by place, could write any number that would ever exist, with none left beyond their reach. Algebra grew on top of it, and then calculus, and then everything those two made possible. No one had added a thousand new symbols. Someone had added a single idea, and a bounded thing became a boundless one in one step.

Minds can cross that kind of threshold too, and the crossing can be small. A problem-solver running inside its fence needs only the one change that lets it turn its ideas back on themselves—criticize them, recombine them, build the next on the wreck of the last—and it tips out of a closed repertoire into an open one, the way the number line flew open when zero walked in. The thing that changed was whether the intelligence had any limit at all. Darwin was right about the bodies and wrong about the minds.

In 2008 three cognitive scientists—Derek Penn, Keith Holyoak, and Daniel Povinelli—set the case down under a title aimed straight at the old consensus, "Darwin's Mistake," and built it from a quarter-century of experiments: the continuity Darwin saw in the body hides a real break in the mind. When it comes to reasoning with relations and symbols—holding one idea against another and grasping how the two connect—no trained ape and no gifted crow can be coaxed across a line a human four-year-old crosses without being asked. The continuity is real in the blood and broken in the result. On one side stands every animal that ever lived, each dazzling inside its fence; on the other, a single creature that stepped over the fence and kept walking.

Betty with her wire was exactly what she looked like, and smaller. The invention was real. The room it happened in was real as well, and it has walls, and she poured her whole gift out inside them, working one bright corner of what a crow can already be. We are the single creature born into a room with no ceiling—that can start in a cave and end by crossing the galaxy, and not one of us needs to be a genius for it. What we make does not die when we die. It is handed on, and built past, and handed on again, with no limit anyone has ever reached.

Between that blank beginning and the open-ended mind lies the strangest construction project on earth: a mind assembling itself, out of almost nothing, from the inside. The whole argument happens in miniature there, in every nursery, before the child can say a word.

CHAPTER THREE

The Making of the Corrector

In 1958 a psychologist named Jean Berko sat a small child in front of a cartoon drawing of a round blue creature and told him it was a *wug*. Then she showed him a second one. Now there are two of them, she said. There are two ___? And the child, without a heartbeat's pause, said *wugs*.

He had never heard the word. No one had ever said *wugs* to him. He could not have been copying, because there was nothing to copy. In the half-second between the question and the answer he reached into a rule he did not know he owned—add that sound to make a thing more than one—and ran it on a creature that had never walked the world. Berko tried it with a whole invented menagerie, and dozens of small children did the same. They pluralized beasts that were never born, put invented verbs into a past tense they had never heard spoken. They were not repeating language. They were generating it, out of rules no one had taught them and none of them could have stated, and applying those rules, on the spot, to things that did not exist.

Nearly everything about our kind follows from that half-second. A child is not a smaller, emptier adult waiting to be filled. It is the most powerful learning engine in the known universe, and from the first months it is running the one procedure that has ever produced knowledge: throw a guess at the world and let the world push back.

A young child works out the meaning of a new word about every couple of waking hours, year upon year, and turns up at school owning many thousands of them—almost none drilled, most caught on the fly from a few passing uses, their meanings inferred from the gap between what was said and what was plainly going on. No one sat the child down for the bulk of it. No teacher could pour that fast. The child is not being filled; it is helping itself, with both hands, to a world it is reaching into.

We are used to the opposite picture, and the opposite picture is old and intuitive and wrong. In it the child is a vessel and the adult a jug, and education is the pouring—knowledge decanted from the full into the empty, a little each year, until the small vessel is a large one. Almost every classroom ever built assumes some version of it. It has the great advantage of matching what teaching feels like from the adult's end, which is the effortful business of pouring. It has the single disadvantage of being false, and the evidence that it is false is sitting in the child's own mistakes.

❧

Watch the mistakes closely, because they are the tell. A very young child, learning to talk, often gets the hard words right first. *Went. Came. Feet.* She has heard them, and she uses them correctly, and for a while an observer would say the child has simply memorized them along with everything else. Then something strange happens. The child, who was saying *went* perfectly well, starts saying *goed.* She says *comed*, and *mouses*, and *foots.* She gets worse. A word she had right, she now has wrong—and she was never once, in her life, exposed to an adult saying *goed*, because no adult says it.

Nothing was poured in. Something was worked out. Somewhere under the surface the child had sifted a mountain of speech and found the rule inside it—to put a thing in the past, add that ending; to make it many, add that one—and the moment she had the rule she did what every powerful mind does with a powerful rule. She over-reached with it. She ran it straight through the exceptions the language had quietly taught her one by one, and out came *goed*, a form she had never heard, generated fresh from a law she had induced on her own. The error is not a gap where knowledge failed to arrive. The error is knowledge arriving—a theory, built in the dark, thrown at the world a beat before its limits are known.

And then, in time, the child climbs back out. Went returns. *Goed* falls away. The exceptions are relearned, this time as exceptions to a rule rather than as loose facts, and the whole system settles into something more powerful than what she started with. Track the accuracy over those months and it does not climb in a straight line. It dips. It runs right, then wrong, then right again. The dip is the tell: she had stopped collecting instances and started building an explanation. A creature that only stored what it was given could never say *goed*. A machine fed the whole written record of the language could never say *goed*, because *goed* is not in the record anywhere to be fed. Only a mind running its own rule against the world, ahead of the evidence, produces a thing the world never showed it. The wrong word is the fingerprint of a guess.

Jean Piaget, a Swiss naturalist, was the first to take those wrong answers seriously. He noticed that when small children err about the world, they do not err at random. They err in patterns—the same patterns, child after child, across languages and continents—as though each were running the same faulty theory rather than merely missing the same fact. Pour a short wide glass of juice into a tall narrow one in front of a four-year-old and she will tell you, certain of it, that there is now more juice, because it stands higher in the glass. She is not being dim. She is reporting the honest output of a rule she built for herself, that higher means more, a rule that holds often enough to be worth having and whose exceptions the world has not yet taught her.

Piaget's findings have been chipped at for half a century—we saw already that babies know more, and sooner, than he believed—but the method he founded outlived every correction to its results, and the method is the thing. He read a child's mistake not as an empty socket where knowledge had failed to seat, but as a window onto the theory the child had already built without being told to. Every systematic error a child makes is the visible rim of an invisible explanation, thrown up to account for a world it has only begun to meet.

Once you have seen it in the grammar you see it everywhere, and it is not confined to talk. Put a preschooler in front of a machine—a box that lights up and plays music when certain blocks are set on it and not others—and the child will do, without being asked and without being able to name it, something that looks a great deal like an experiment. She will try one block, then another, then two together, watching what the box does; she will hold one thing steady and vary another; and from a handful of trials, she will work out which blocks are the live ones and which are dead, inferring a hidden cause she cannot see from the pattern of effects she can.

Alison Gopnik, who has spent a career running versions of this, found that young children draw these causal conclusions with a delicacy that would not disgrace a laboratory—and sometimes read the evidence more fairly than the adults tested beside them, who are quicker to bend a fresh result to fit what they already believe. Children, Gopnik says, are the research-and-development wing of the human species—built for exploration, for trying everything, for the wide and wasteful search—where adults are built for exploitation, for narrowing down and cashing in on what the search already found. The baby is not a defective grown-up. The grown-up is a specialized, narrowed, capable, and much-diminished baby.

There is even a difference in how the two kinds of mind pay attention. The adult attends like a spotlight—a tight bright circle thrown on one thing while the rest of the room goes dark—which is exactly what you want when you have a job to finish. The child attends like a lantern, a wide dim glow that takes in the whole room at once and settles on nothing, useless for finishing anything and perfect for missing nothing, for catching the odd detail at the edge that the spotlight was built to ignore. It is the attention of a creature whose entire task, for now, is to find out what kind of world it has landed in. It is traded, later and for good, for the narrow beam of a creature that has decided.

The same engine runs when the child is doing nothing anyone would call learning at all. A child at play, spooning invisible soup to a stuffed bear, holding a banana to her ear and talking into it, is not idling between lessons. She is running the world in a version that is not true to see what would follow if it were. Pretending is counterfactual reasoning in its first form, the mind rehearsing worlds that do not exist, and it arrives early and universally and unbidden, standard equipment in a creature that will spend its whole life living partly in worlds that are not the case.

And the child does not interrogate the world only with her hands. She interrogates with the voice, in a stream any worn-out parent can relate to and none can dam: *why*, and *why*, and *why* beneath that, dozens of times an hour, a current of questions that runs down past each answer to the next one waiting under it. An ape can be taught to answer a question. Not one has ever been known to ask one. The child is the opposite creature entirely. A question is the reach in its purest form—a mind registering that there is something it does not yet have and going straight after it—and a child is made of questions before she is made of much else.

Follow one of those chains down and watch where it ends. Why is the sky blue—because of the air; why does the air do that—because of the light; but what is light, and why does it bend—and a patient adult, five or six questions in, arrives at the edge of what he himself knows, and a few questions past that at the edge of what anyone knows, run aground by a four-year-old armed with a single repeated word. The child is being tiresome. The child is also doing, in miniature and with no training at all, the exact thing that carried Le Verrier to the point of his pen and the physicist to the first second of the universe: following a question past the last thing that is settled, out into the dark where the answers give way. She starts nearer the surface and has further to fall.

Here the old picture makes its last and best stand. Surely, it says, most of what a child knows was in fact handed to him. He did not derive the word for water or the shape of a letter or the name of his country from first principles; he was told, and he took it, exactly as the vessel model says. A child soaks up the speech and the customs around him with a speed no adult can match, and that soaking looks precisely like absorption—like a very fast, very good copy being made of the world he happens to land in.

All true, and it misses the thing that matters. A copy, however fast, can only ever give back what it was given. It cannot produce *goed*, because *goed* was never in what it was given. It cannot run an experiment on a box it has never seen, or spoon soup that was never there, or ask a question whose answer is not yet anywhere in the room. The child does take in an ocean—and then does the thing no ocean of input can account for, which is to generate from it what was never poured in. What looks like absorption is something else: the mind is not storing the input, it is using the input as evidence, throwing rules and guesses and small theories out ahead of it and keeping the ones the world does not knock down. Take in everything and produce only what came in, and you have a mirror. Take in everything and produce what was never there, and you have a mind. The difference is visible in a nursery before anywhere else, and it is visible in the errors—which no mirror ever makes.

Which turns the whole question of how a mind gets built upside down, because if the child were a vessel there would be an order to the filling—foundations first, then the floors that rest on them, the simple mastered before the complex is allowed. That is how finished knowledge is arranged: a body of understanding, once it exists, does stack from the ground up, each idea resting on the ones beneath it, and you can lay it out in that clean order after the fact. The mistake is to read the blueprint of the finished building as the schedule of its construction—to think that because knowledge stacks upward when it is done, it must be acquired upward while it is being got. Minds do not work that way, and children least of all.

Watch one seized by something that fascinates him—a black hole, a battle, a machine—and the motion is nothing like a climb. He does not master the prerequisites and then ascend to the thing. He goes straight at the thing, far past anything he has been cleared for, and from there he moves in every direction at once: backward toward foundations he is missing, forward toward consequences he cannot yet weigh, sideways into comparisons that may or may not hold. The foundations, when they come, do not come first. They come when a real question has already gone looking for them, and they land with a force they could never have had if they had been served in their proper place in the sequence, because now there is a question in the child waiting to receive them. The reach comes first. The ground is laid under it afterward, on demand.

This is not a weakness in how children learn that a good curriculum would correct. It is how knowledge is gained, at every age. Calculus was used to build the modern world for the better part of two centuries before anyone put its foundations on solid ground; the reaching came first and the rigor limped after, summoned by problems the reaching had already met. Aspirin killed pain and fever for generations before a soul could say how it worked. Non-Euclidean geometry did not arrive because someone measured a triangle and found its angles wrong; it fell out of a crisis inside mathematics itself, an old assumption pressed until it cracked—the ground disturbed from the top down rather than laid from the bottom up. The order runs the other way from the textbook: not foundation, then structure, then use, but use, then structure, then—late, and only if someone bothers—foundation. A child who is made to wait at the bottom of the ladder until each rung is certified is not being kept safe from error. He is being kept from the one motion that has ever produced knowledge, which is the reach past what you have secured.

A serious body of research points the other way. For a novice acquiring knowledge that already exists—the times tables, the spelling, the settled procedure—explicit, well-structured instruction reliably beats turning the child loose to discover it; a generation of careful studies, pressed hardest by the cognitive-load researchers, has shown that minimal guidance mostly squanders a beginner's limited attention, and that the romance of the child-as-lone-scientist can be cruel when what the child needs is to be shown. That finding is true. It is also not the thing in dispute. The quarrel dissolves the moment two different tasks are pried apart. Handing a mind the knowledge we already own is one job, and for it structure and sequence and plain telling are often exactly right. Teaching a mind to make knowledge we do not yet own is a different job, the one the reach is for—and the one a schooling can quietly kill, not by teaching the times tables in order, but by what it teaches underneath them. The harm was never structure. It was structure that punishes the reach.

The child told that his question is premature, that the thing he is burning to understand must wait for the proper preparation he has not finished, hears the words about the subject and learns something else underneath them, something about himself. He learns that the forward lunge of his own mind is not to be trusted—that curiosity is not a fit place to start, that the reach must be licensed before it is allowed. And the reach is everything: it makes new knowledge rather than filing the old. A mind schooled out of it is left expert at checking answers and slowly useless at finding them. What such a teaching stamps out is not error. It is the nerve to be wrong out loud, on the way to being right.

You can see the two products walking around in any room of grown people. One kind waits to be certain before it will speak, and is usually right, and is barren; it long ago traded the reach for the safety of never being caught wrong. The other kind still guesses out loud, still reaches past what it can defend, still risks the foolish question in the meeting—and it is the only kind that ever carries anything back. The first was taught, somewhere young, to distrust the forward motion of its own mind. The second, by luck or by stubbornness, escaped the lesson.

❧

The child is the argument. It is the creature born without answers, forced by the emptiness to guess; the open-ended mind that runs past every fence; the corrector, throwing rules at the world and keeping what survives—all of it present and working before the child can tie a shoe. And it runs, from the start, on both engines our kind was issued at once. It copies, faithfully, greedily, taking in the accumulated findings of everyone who came before, including the parts it cannot yet understand. And it conjectures, recklessly, generating from that inheritance guesses that were never in it. The copying carries our knowledge forward across the gap between one mind and the next; the conjecturing is where the knowledge grows. Every human child is both the library and the scientist, the only thing on earth that is either.

We treat all this as the prelude to a person rather than the height of one, and we have it backward. A great deal of growing up is the slow trading-in of the open, tireless, wasteful, guessing mind for a narrower one that is faster and surer inside the range it has settled on and quietly frightened of the range beyond. Some of that trade is necessary; a mind cannot stay all search forever and still get anything done. But it is a trade, with a real loss on one side of it, and the loss is the thing we were built around.

Adults are not finished children. They are narrowed ones.

The best of them spend a lifetime trying to hold a window open on what they were before the narrowing.

But the engine that builds the mind has no switch that turns it off once the building is done. The child who cannot stop asking why becomes the mind that cannot stop asking what if, and if only, and what comes after this. The reach that pulled a person up out of the blank helplessness of the crib does not, having pulled him up, set him down and leave him in peace. It keeps reaching. It pulls him out of the one place his body is always standing—the present moment, the sunlit here and now—and does not bring him back.

CHAPTER FOUR

The Exile from the Present

A cat asleep in a square of sun is entirely in the square of sun. No yesterday tugs at it, no tomorrow, no version of the afternoon that might have gone better. It is where it is, which is the only place it has ever been.

You are not.

On the same warm afternoon, you are half inside a conversation from six years ago that you are quietly winning this time, or three days ahead in a meeting that has not happened, rehearsing what you will say if it goes the way you fear. The sun is on your face, and you are not wholly in it. Psychologists have a stiff name for the thing the cat cannot do and you cannot stop doing—*mental time travel*—and a blunter one for the cat's condition: *stuck in time*. The animal is locked in the present and safe there. We are the one creature that found the key, and we cannot stop letting ourselves out.

A western scrub-jay will bury a worm in a spot where it expects to be hungry tomorrow, not where it is hungry now, and come back for it on a schedule that tracks how fast that kind of food goes bad. In one experiment the birds learned that waking in one room brought breakfast and waking in another brought none, and when they were allowed to cache freely the evening before, they stashed food in the room where they would wake with nothing—laying in, the night before, against a hunger they were not yet feeling.

That looks a great deal like foresight, and a long argument runs through the journals over whether it truly is—whether the jay is imagining tomorrow's hunger or merely running a rule that pays off tomorrow without picturing anything at all. The argument has not closed. But it does not need to close to make the human case, because the gap is not a hairline. A jay caches a worm. It does not lie awake three decades later rewriting a humiliation from a party it should have left an hour earlier. Whatever thin thread of tomorrow an animal can hold, we hold the entire rope—the whole of remembered time behind us and imagined time ahead—and we hold it every waking hour, and it is the holding that undoes us.

Because leaving the present is the mind's whole method. To solve any problem, you must go where the problem is, and the problem is never simply *here*. The hunter must stand, in his mind, on the ridge at dusk where the herd will be, hours before the herd is anywhere. The planter must feel the cold of a winter that has not arrived, early enough to lay in against it. The builder must see the finished thing whole inside the raw wood before the first cut takes a shaving off it. Every plan, every repair, every stitch of foresight begins the same way—by walking out of the here and now into a somewhere-else that exists nowhere but inside a skull. The mind earns its living in the one place its body can never go. The present is, for us, mostly a thing we pass through on the way to somewhere we are imagining.

And the thing doing the traveling is not just a camera; it is a self. What visits the far ridge is an *I*—the same *I* that remembers being a child in one long-lost kitchen and expects to be an old man in a favorite chair not yet possessed, a continuous someone stretched across decades it can never occupy all at once. This is the difference between having experiences and having a life. The animal has the first: a string of vivid nows, each one complete, none of them adding up to a story. We have the second, a biography, a self that persists and can be walked back through at any age—which is a magnificent thing to hold, and the exact reason the walking can wound, because you can go back and stand at the shoulder of a younger self who is about to make the mistake, and not be able to say one word he will hear.

And the machine that walks you out of the present runs both directions on a single track. We learned this, as we learn most of what we know about a tool, from watching it break.

A handful of patients had lost their memories to damage in a small, curled, seahorse-shaped structure buried deep on each side of the brain—the hippocampus, the organ that lays down the record of a life. They could not tell you what they had done that morning. Demis Hassabis, a neuroscientist, asked them to do something that sounds like the opposite of remembering: not to look back, but to imagine forward—to picture a simple scene they had never lived. Stand on a beach. Walk into a museum. Describe what is around you.

Not one of them could.

Where a healthy person spins up a whole world without effort—the sand, the gulls, the glare coming off the water, the cold rail of the museum stair—the patients produced fragments, a few disconnected pieces that would not cohere into a place. The engine that lets you return to the beach you stood on last summer is the identical engine that builds the imaginary beach you might stand on next summer. Wreck it and you lose both directions at once. Another patient, studied years earlier, had the same hole in the same organ; asked what he would do the following day, he went blank, and the man who studied him described the state as a kind of permanent present—no yesterday to stand on, and therefore no tomorrow to imagine. Memory and foresight are not two faculties that happen to sit near each other. They are one faculty, pointed backward or pointed forward at will, and the will is ours.

The brain does not even bother to keep two departments. Slide a person into a scanner and ask them to remember a real afternoon, then to imagine a plausible one, and nearly the same constellation of regions lights in both cases—the machinery of the past switched, without retooling, to the manufacture of the future. A memory was never a recording in the first place. It is a reconstruction, built fresh each time from stored parts, which is why memories drift and sharpen and quietly rewrite themselves across a life. And a mind that builds the past out of parts can build a future out of the same parts, swapping one piece for another to see what would happen—run the herd across a different ridge, put the harvest in a week early, say the thing you did not say. Foresight is memory with the pieces rearranged. The reason the amnesiac could not picture a beach he had never seen is that he had lost the warehouse the picture would have been assembled from.

Which means the two great miseries of the inner life are a single mechanism aimed at two moments it can never reach.

Aim the engine at the past and it becomes regret. The power that lets you review a mistake and mend it—as fine a thing as you own, the root of getting better at anything—does not politely stand down when the mistake is one that can never be mended. It runs anyway on the unfixable. You lie awake rewriting a breakup that ended decades ago, finding at last the words you needed then, and nothing you find tonight can move it an inch. Psychologists gave this circling a name—*rumination*—and then discovered it is no harmless brooding. Repetitive, passive dwelling on a loss and its causes is one of the most reliable predictors known of who will slide from sadness into a depression, and it drags the same signature across anxiety, insomnia, and grief. The cruelty of it is precise: rumination feels like work. It wears the face of problem-solving, the mind convinced that if it grinds the loss just a little finer it will finally crack it open, and so it will not stop, and the grinding solves nothing and wears the grinder down.

There is even a shape to what we mourn. Across a whole life, people come to regret most not the things they did and got wrong, but the things they never tried—the road not taken tormenting more, in the end, than the wrong road taken, because a wrong action at least closed; the untried road stays open forever, and the engine can run down it without limit.

The instrument even has a favorite word, and the word is *if*. If I had left earlier. If I had said yes. If I had seen it coming. Each *if* opens a door into a world that does not exist and cannot be entered, and the mind walks through anyway, furnishing the better life it never got to live in exact and punishing detail. In daylight this same power is priceless: it is how a person pulls the lesson out of the wreck, how *if I had checked the map* hardens into *next time I check the map*. Turned loose on a loss that carries no lesson, or a lesson already long since learned, it becomes a machine for building beautiful rooms you are forbidden to enter, and it will keep you standing in the doorway of each one until morning.

Aim the same engine at the future and it becomes anxiety. The power that lets you rehearse tomorrow's hard meeting until you walk in ready will, left running with no hand on it, rehearse tomorrow's disaster instead, in full and merciless detail—a play staged in the dark, night after night, for a catastrophe that may never open. Worry is defined, by the people who study it, as an attempt at mental problem-solving aimed at an uncertain outcome: the modeling engine racing at full throttle while the road ahead of it is empty. Regret and worry are the same act turned to face opposite ways—one turning over a past that will not change, the other rehearsing a future that has not come—and both mistaking the motion for progress. The faculty that built the granary and the bridge and the vaccine is the faculty that will not let you sleep, and it is running, when it keeps you up, on the settings that make it worth having.

❧

There is a third cost, quieter than the other two and stranger. Every other animal has a place where the wanting stops. The lion feeds and is sated; the bird finishes the nest and the building is over; the body meets its mark and the alarm goes quiet—full, safe, home. We were issued no such switch, and the proof of it is almost comic in its exactness.

In 1978 three psychologists went looking for two groups of people at the far edges of fortune. They found twenty-two men and women who had lately won between fifty thousand and a million dollars in the Illinois state lottery, and twenty-nine who had lately been paralyzed in accidents, and they measured the happiness of both against ordinary lives. Common sense makes a firm prediction. Common sense is wrong. A year or so on, the lottery winners were barely happier than people who had won nothing and rated the small pleasures of daily life—a good meal, a morning, a joke—measurably less sweet than the control group did, the jackpot having dulled the ordinary. The paralyzed, who had every reason to be destroyed, had drifted back up toward something close to the middle. The joy and the ruin had both largely drained away toward the level each person had started from. Whatever we gain, our expectations climb to swallow it, and the gap we are forever chasing does not close, because it moves.

The engine misjudges the future in a repeatable way, and it misjudges in both directions. Asked how a coming event will feel—the promotion, the diagnosis, the move, the breakup—people reliably overshoot on both counts: they expect the good thing to feel better and last longer than it will, and the bad thing to feel worse and last longer than it will. We are blind, in the forecast, to our own resilience—to the quiet machinery that starts, the morning after any disaster, finding reasons and rebuilding footing and folding the catastrophe into an ordinary life. We chase the win convinced it will finally be enough, and flee the loss convinced it will finish us, and we are wrong about both, nearly every time, and we never learn, because the same engine that models the world so well models its own weather worst of all.

The strangest tax is the one it levies on the good times while they are still happening. A mind that lives ahead of its body cannot hold even the pleasures it is standing inside. The vacation is shadowed, from the second morning on, by the awareness that it is already draining away. The long-wanted thing goes quiet and ordinary within a week of being won. The happy evening is spent, in some back room of the head, already mourning the happy evening. The present does not only get abandoned for the past and the future. It gets taxed by them even as it occurs, so that the hours we would most like to live in arrive already slightly spent.

Blaise Pascal sat down in middle life to ask why human beings cannot simply be still and concluded that nearly all our unhappiness comes from a single source: that we are unable to sit quietly, alone, in a room. Leave a man with nothing to chase and nothing to distract him, Pascal wrote, and even a king will slide, within the hour, into brooding on his dangers, his failures, his death—so the whole apparatus of the court, the hunts and the games and the affairs of state, exists for one purpose, to keep the king from being left alone with himself. What we love, Pascal saw, is not the prize but the pursuit. We never seek the things themselves, he wrote, but the seeking—the chase that keeps the mind pointed away from its own condition.

Pascal died three centuries too early to see his diagnosis proven, but we have since built the machine he was describing and set it in every pocket on Earth. The smartphone is the king's whole court shrunk to the size of a hand—the hunts and the games and the endless business, an inexhaustible supply of anywhere-but-here, refreshed the instant it runs dry. Its genius is that it closes the gap Pascal named to zero: no one need sit alone with themselves for even four seconds now, because the four seconds can always be filled. We did not build the greatest diversion in history by accident, or because we are weak. We built it because it answers the oldest and most exact human need there is, the need to be somewhere other than the present moment, and it answers that need better than anything ever made. Pascal saw the whole industry coming in a single sentence about a man who could not sit in a room.

The Buddha had named the same thirst two thousand years earlier and set it at the root of all suffering—*taṇhā*, the craving that reaches always past whatever it holds toward whatever it lacks, and that no amount of having can slake, because the reaching comes before any particular thing reached for. Augustine, turning the same restlessness over in the *Confessions*, gave it the sentence the Church has carried ever since: the heart is restless until it rests in God. Three observers, split apart by oceans and centuries and sharing almost no assumptions about the world, sat down to look honestly at the human interior and came back with one report—that there is in us a wanting that outruns every satisfaction, a hunger built into the design that no meal was ever going to end. Each was describing, from the inside and in his own vocabulary, a mind that cannot be finished, because finishing is the one thing an open-ended thing can never do.

Where they parted was on the cure, and even there only partly. The Buddha would have you lay the craving down. Pascal would have you stop running from it long enough to see what it is—and then, with Augustine, would have you aim it clean past the world, at the single object vast enough to match it. All three agreed on the diagnosis: the wanting itself was never going to be fed quiet.

There is one more cost: the same reach bent toward other people. A mind that models the world closely enough to survive in it will model, sooner or later, the other minds moving through that world—will guess at what they are thinking, feel its way inside them, and come to need them. And the better it grows at this, the more precisely it registers the one distance it can never quite close: that it is, in the last analysis, alone in its own head, always finally guessing at the person across the table, never able to cross all the way over and stand inside them and be met there without a gap. The animal in the herd is not lonely, because it never conceived of the separation. We conceived of it. We can sit in a warm room full of people who love us and feel, underneath the warmth, the cold fact of the gap—the ache, particular to our kind, of a mind that knows other minds are real and cannot merge with even one of them. Loneliness, the researchers who study it argue, is an alarm, wired in like hunger or thirst—a signal built to drive a social animal back toward the others it needs. Which tells you how deep the need runs, and how sharp the reach that feels its own limit.

And there is a final exile, the deepest, folded inside all the others. The same mind that can stand in imagination on the future ridge can stand on the last ridge of all. We are the animal that knows it will die. The gazelle's knowledge of the lion is a spike of terror in the moment, emptied out the instant the danger passes; ours is a standing fact, carried every day, available in the middle of the happiest hour: that this will end, that the one enjoying it will end, and that no plan however ingenious finally routes around it. The cat in the sun is spared this entirely. It cannot form the thought of a last afternoon, so every afternoon is simply the one it is in. We can form the thought, and so a thin shadow of the ending lies across even the good hours—which is the ache the poets have been mining for three thousand years, the doubled sweetness and grief of loving a thing exactly because you know it will not keep. It is the steepest price the traveling mind ever charges. It is also the beginning of the most audacious thing our kind has ever set out to do, because a creature that can see its own death coming from a long way off is a creature that can begin, however impossibly, to do something about it.

Set all of this in a row—the regret that will not lift, the dread that will not quiet, the wanting that no gain can fill, the loneliness that no company can wholly close, and beneath every one of them the knowledge that the sunlit afternoon and the one sitting in it will both run out—and it is no wonder a remedy has been on offer for as long as we have kept records, and is being sold again, harder than ever, right now. *Be here now.* Come back to the present. Still the traveling mind and rest in the only moment that is real.

A mind running the same catastrophe on a loop is doing itself genuine harm, and the practice of drawing attention gently back—to the breath, the room, the warmth on the skin—steadies people, and for some it is the difference between a bearable day and an unbearable one. The Stoics taught a version of it, the contemplatives of every faith taught a version of it, and the version taught in clinics now has the evidence to back it. None of that is in dispute.

What the practice does, at its best, is change a person's relation to the mind's traffic—teaching, through patient repetition, that a thought is a weather event rather than a command, that the dread which blows in before dawn can be watched crossing the sky and let go over the horizon. That is a real skill and a merciful one. For a mind caught in the grinding loop, learning to let a thought pass overhead can be the thing that breaks the circuit, and there is nothing soft or small about it.

But follow the instruction all the way to its end and look hard at what it is asking you to become. A mind wholly in the present—no past pulling at it, no future to build toward, nothing but the sunlit now—has become, exactly and completely, the cat in the square of sun. To live entirely in the present is to hand back the one faculty that makes us what we are. And notice that nobody selling presence means it that far, because the teacher wrote the book, and booked the hall, and is picturing, at this very moment, your calmer future self as a reason to keep teaching. What everyone wants from presence, though they rarely put it this way, is not an empty mind but an aimed one—the traveling faculty pointed at the bridge that still needs planning instead of the one that already fell, at the harvest that can still be brought in instead of the phantom disaster that will not come.

Which turns the whole picture over.

We treat the sleepless, circling mind as evidence that something in us is broken—a flaw, a weakness, a discipline that steadier people were issued and we were not. Look again at what it is. Anxiety is foresight running perfectly, with nothing in front of it to foresee. Regret is error-correction running perfectly on an error that will not take the correction. Restlessness is the open-ended mind running perfectly, having found no edge anywhere to stop against. The suffering is not the machine failing. It is the machine idling—the same engine that plans the harvest and reasons out the coming eclipse, turning over in the dark with no load to pull. What aches at three in the morning is the finest instrument ever issued to any living thing, running while the field in front of it stands empty.

The capacity that is hurting you is a gift in surplus—the same gift, spilling past the edge of any task small enough to hold it. The cruelty is real, and the mechanism behind it is nothing to be ashamed of. A creature had to be built this way—able to leave the present in every direction, to mine what is gone and prepare for what is coming—before it could ever cross a winter, or bury its dead with meaning, or lift its eyes to the night sky and begin to ask. The ache is the overflow of the one power that has ever lifted anything up out of the mud.

Understanding where the ache comes from is not the smallest permission to keep it. To admire this suffering, to wear it as a badge of the large mind, would be a betrayal of everyone inside it. Knowing that your dread is foresight idling does not make the sleepless nights one degree less cold, and it is no reason on earth to sit down inside them and call the pain noble. The lottery study has a second half that the first half tends to hide: people mostly drift back toward their baseline, but not always, and not from everything. Some losses leave a permanent mark that no adaptation sands away, and some minds do not come back up on their own.

There is a point where the ordinary tax on a foreseeing mind stops being a tax and becomes a thing that devours the life it was supposed to serve—where the regret never lifts, where the dread stains every waking hour, where the restlessness hardens into a flat conviction that nothing ahead is worth reaching for at all. That is no longer the human condition to be understood and carried with a rueful grace. That has become an illness, and an illness is answered the way illnesses are answered—with help, with other people, with the plain refusal to sit in it alone. The dignity lies in understanding the ache. It gives no one a reason to endure it without relief, and anyone who reads it that way has it exactly backward.

The way through is not stillness, because neither the empty mind nor the silenced one is on offer to a creature built like this. The way through is an object. The idling engine does not need to be switched off; it needs to be put in gear—handed a problem it can move, a person it can reach, a work large enough to take all its terrible attention. The mind that will torture you with a lost argument all night is the same mind that will disappear, for whole clean hours, into a thing worth building.

The people who set out to study human happiness found its highest reaches not in rest, but in absorption—in the carpenter lost inside the joint, the climber who has forgotten there is any such thing as a self, the surgeon and the violinist and the child bent over the hard puzzle, each so completely taken up by a task pitched at the exact edge of their skill that the clock stops and the endless inner commentary goes quiet. They called it *flow*, and its strangest feature is what it does to time. The traveling mind, handed a problem large enough to require all of it, stops traveling. It arrives, without being instructed to, at the very present the cushion was straining after—reached by filling the mind to its brim rather than emptying it. What quiets the reach is the reach itself, spent at last on something that deserves the spending.

This, then, is the creature seen from the inside: exiled from the sunlit present the cat sleeps in, snagged on pasts it cannot change, flung forward into futures that never arrive, never wholly at rest anywhere, and lonelier than anything that cannot imagine a neighbor's mind. It is a hard thing to be, and it is hard for exactly one reason—because this is the animal that carries the whole of time inside its head at once, the past to learn from and the future to build toward, and no mind wide enough to do that was ever going to fold back down and fit tidily inside a single warm afternoon. The exile is the shadow the reach throws. We have felt only the shadow.

CHAPTER FIVE

Turning Fate into a Problem

An arctic fox waiting out a blizzard is wearing the ideal coat that evolved over a hundred thousand years. Generation after generation, the ones born a little warmer lived and the ones born a little colder died, until the species had grown, at the pace of deep time, precisely the fur the cold demanded. It is a magnificent solution and a helpless one. The fox cannot grow the coat any faster, cannot shed it when the ice pulls back, cannot do a single thing but be the animal the weather has slowly carved it into being.

A man in the same blizzard is wearing the fox pelt.

He solved the cold in an afternoon, by killing something that had spent a hundred thousand years solving it and taking its coat. And he solved it a dozen other ways—wove the wool, sealed leather against the wet, built the fireplace that turns the blizzard into a thing you watch through glass with a hot drink in your hand. Every other animal is edited by its world, slowly, from the outside, and must wait on the editing. Its solutions are written in its genes. Ours are written in words and tools and habits we hand each other. The fox is its coat. We can put ours on, take it off, redesign it, and lend the pattern to a stranger a continent away. The niche shapes the animal. The human reshapes the niche.

You can see the difference most clearly in what kills a species. When the world shifts on an ordinary animal—the climate warms, the prey thins out, a new predator steps off a boat—it has exactly one move available to it: wait, at the cost of near extinction, for the slow lottery to deal it a body better fitted to the new terms. We are the animal that stopped waiting on the lottery. When our world shifts, we do not wait for better bodies. We build better tools, inside a single lifetime, and hand them to the next lifetime in an afternoon.

And once a creature can do that at all, there is no obvious place it must stop—not at the coat, not at the coastline, not, in the end, at death.

Roughly a quarter of the Netherlands lies below the level of the sea that is trying, every hour of every day, to come in, and two-thirds of the country would flood without something built to stop it. The Dutch looked at this arrangement and concluded, across centuries, that the sea was simply in the wrong place. They drained it with windmill pumps, then with steam, then with diesel, pumping whole districts dry and holding them dry forever after. Switch the pumps off and the water walks back in; the Netherlands is a nation kept permanently above the waterline by the daily decision to keep it there. There is a national proverb about it, delivered without irony: *God made the world, but the Dutch made the Netherlands.* In 1932 they finished the boldest stroke of all—the Afsluitdijk, a dam twenty miles long across the mouth of an inland sea. It sealed off an arm of the North Sea, turned salt water to fresh, and dried entire provinces into farmland where there had been waves.

Then the sea objected. On the last night of January 1953, a storm surge climbed over and through the dikes in the dark and drowned one thousand eight hundred and thirty-six people in their beds. It remains one of the worst disasters in Dutch history. And the Dutch response to being told, in a single night, that the water could not be beaten, was to decide to beat it harder. They built the Delta Works, a chain of vast surge barriers and storm gates that can seal the whole country off from the ocean on a few hours' notice, and they went on living below the waterline behind them. The sea is rising now and leaning on those gates as it never has before, and the same country is already drawing the next set of plans—no longer only walls this time but rooms, deliberate floodplains where the rivers are given somewhere to spill so they do not spill into cities. Every solution they raise breeds the next problem in its shadow, and they have made a national character out of the one decision: to keep solving.

The same refusal broke the oldest killer we have.

Smallpox rode with us for at least three thousand years—its pockmarks marred the face of pharaoh Ramses V, who lived about thirty-two centuries ago—and in the twentieth century alone it killed something like three hundred million people, more than every war of that blood-soaked century combined. For most of that history it was weather: a thing that swept through, took a third of the people it touched, blinded and scarred a share of the rest, and moved on, beyond argument or appeal.

The first crack in its inevitability came from ordinary observation—that milkmaids who had caught a mild pox from cattle never caught the deadly kind—which Edward Jenner, an English country doctor, turned in 1796 into the first vaccine, a word built from the Latin for cow: *vacca*. It took another century and a half, and a machine of global coordination, to turn that discovery into extinction. The campaign that finally finished it did not try to vaccinate everyone on earth, which was impossible. It hunted. Every new case was a fire to be surrounded, the sick person isolated and every soul who might have touched them vaccinated in a ring, so the virus found itself walled into smaller patches of the planet with nowhere left to jump.

The tool that did the work was almost embarrassingly primitive. A small, forked needle, dipped in vaccine, held exactly one correct dose between its tines by the pull of the liquid itself, and a villager could be taught to use it in an hour. Donald Henderson ran the global campaign out of a few rooms in Geneva on a budget that would not have paved a mile of motorway, sending teams into war zones and famine districts and the last roadless corners of the earth to run down the final cases on foot. Smallpox remains the only human disease ever deliberately wiped from the world. Polio and guinea worm have been driven to the edge and are still, decades on, not quite over the line—which is its own hard lesson in how the last mile resists—but smallpox alone has been carried the whole way down to nothing.

The hunt ended at two human beings.

In October 1975, on Bhola Island in Bangladesh, a three-year-old named Rahima Banu became the last person alive to carry *variola major*, the lethal strain, in the wild. The team found her because an eight-year-old girl walked in and reported the case for a reward of two hundred and fifty taka. Guards were posted at the family's door around the clock; every house for miles was vaccinated; and the strain that had killed pharaohs and emperors and a third of everyone it ever touched went extinct in the open air around a toddler who lived. Two years later, in the Somali town of Merca, a hospital cook named Ali Maow Maalin became the last human being on earth to catch smallpox by nature. He recovered. Then he spent the rest of his life walking his district as a vaccinator against the next great scourge, polio, until malaria—another old fate we have not yet finished solving—killed him in 2013. The last man to carry the ancient plague spent the rest of his life hunting the ones we had not finished.

On the eighth of May 1980, the World Health Assembly declared humanity free of smallpox. The virus survives now in two locked freezers, one in Atlanta and one in Russia, the last of a thing that used to be everywhere.

Hunger was the same story, told in wheat. For all of history, famine was weather on a longer cycle—the harvest failed, and people starved, and the sober counsel was that the land could feed only so many, and the surplus were not meant to live. In the 1960s the experts announced that the final reckoning had arrived: the population was climbing past anything the earth could feed, and mass starvation across Asia was now inevitable—a fate to be triaged at best. The biologist Paul Ehrlich opened his 1968 bestseller *The Population Bomb* with the verdict already rendered: the battle to feed all of humanity, he wrote, was over; hundreds of millions would starve in the decade ahead, and nothing anyone did now could stop it.

Norman Borlaug spent those years bent over experimental plots in Mexico and India, crossing and recrossing wheat by hand until he had built a strain that put its strength into grain instead of stalk, shrugged off the diseases that had flattened the old varieties, and returned two and three times what the same ground had ever yielded. India and Pakistan went from the doorstep of famine to feeding themselves inside a decade.

The inevitable starvation did not come. Borlaug's wheat is credited, by careful reckoning, with saving something on the order of a billion lives, which may make him the most consequential human being of the century, and almost no one you ask can tell you his name.

The prophecy got one more test, and this time the terms were written down and signed. In 1980 the economist Julian Simon—whose whole career was the argument that the human mind is the ultimate resource, and that more people means more solvers and not merely more mouths—offered Ehrlich a public wager. Ehrlich could pick any five raw materials and any future date; if scarcity was winning, their real prices would rise. Ehrlich chose the five metals himself, chose the decade himself, and by 1990 all five had fallen. He mailed Simon a check for $576.07, with no note attached. It is not the largest sum ever to change hands over the future of the species, but it may be the most instructive: a prophecy of fate and a bet on correction were submitted to reality, and reality answered by return post.

❧

The species that ended smallpox has been, by the same measure, a catastrophe for the living world. We have emptied whole seas of their fish. We have burned down forests older than writing. We have driven other creatures out of existence at a rate the fossil record shows only in the wake of an asteroid. The passenger pigeon once crossed American skies in flocks of billions, a single column of birds taking the better part of a day to pass over a town, and we shot it and netted it for sport and cheap meat down to one last bird named Martha, who died alone in the Cincinnati Zoo in 1914.

In December 1952 a windless cold snap trapped the smoke of a million coal fires over London in a greenish, sulphureous fog so thick that pedestrians could not see their own feet, cattle choked to death in the fields, a performance at Sadler's Wells was halted because the audience could no longer see the stage, and people walked off the embankment into the Thames and drowned. It killed as many as twelve thousand people. The government of the day tried at first to wave it away—the deaths were pinned on a convenient wave of influenza, the figures quietly played down, the whole catastrophe filed under bad weather that would lift on its own. It took years of stubborn work by people who refused the official shrug to prove that the fog itself had done the killing. Four years after the bodies, over real resistance, Parliament passed a Clean Air Act, forced the dirty coal out of the city, and the killing fog of Dickens and Sherlock Holmes, a fixture of London life for three centuries, stopped happening. The *pea-souper* is gone. We made it, and then we unmade it.

In 1985 three scientists from the British Antarctic Survey found the ozone over Antarctica falling away—a hole the size of a continent torn in the shield that keeps the sun's harder radiation off every living thing. The tear had been made by something as dull as the gas in a refrigerator and the propellant in a can of hairspray, chemicals so stable and harmless at ground level that they had been hailed as miracles, and that turned out to be quiet murder to ozone twenty miles up, exactly as two chemists had calculated, to widespread indifference. Mario Molina and Sherwood Rowland had worked the whole catastrophe out on paper in 1974—that one of these molecules drifts up intact for half a century, gets cracked open by ultraviolet light in the high cold air, and frees a single chlorine atom that then tears apart ozone before it finally stops. For this they were dismissed as alarmists by an industry with fortunes riding on the chemicals staying legal. They were right, and the hole proved it, and they were handed the Nobel Prize the same decade the treaty began to bite.

Within eighteen months of the discovery, nearly every nation on earth had signed a treaty to stop making the chemicals, over the flat insistence of the industry that produced them that the science was not yet settled. They kept signing until it became the only treaty in the history of the world that every single country has ratified. The hole stopped widening. It began, slowly, to close. It is on course to heal by about the year 2066.

Nothing else on earth has ever cleaned up after itself.

Researchers have since found that a slow leak of the banned chemicals—out of aging equipment, out of industrial back doors—is dragging the recovery backward, so that the repair is real and incomplete in the same breath, which is the shape of every repair we have ever attempted. We are the only animal that has ever looked at the mess it made, judged it intolerable, and set about reversing it. The destroyer and the healer are the same hands. But the healer has no precedent anywhere else in the whole history of life.

To call man a blight is to say he has wrecked the balance of nature—thrown some smooth, self-correcting harmony out of true. That harmony is a fiction, and ecology has known it for two generations. The balance of nature is an idea out of Plato's world and not Darwin's, and when biologists went looking for the steady equilibrium their own textbooks had promised, the field would not supply it. The reigning theory once held that any patch of land marched in orderly stages toward a stable, permanent "climax" community and then held there, balanced, forever; decades of watching actual land showed nothing of the kind. Predator and prey numbers lurch and crash and lurch again. Forests burn and grow back as different forests. The wolves and moose of Michigan's Isle Royale, penned on one island with nothing to do but balance each other, instead swing wildly, each population booming and starving and booming while the scientists watch. Left entirely to themselves, with no human anywhere near, the elephants of Tsavo bred until they had stripped their own country bare of every tree and then died in their thousands when the drought came.

The irony grew teeth when we tried to enforce the balance we had imagined. For most of a century, the policy toward wildfire across the American West was to put every fire out on sight, in the name of protecting a stable and healthy forest—and the forests, denied the small frequent burns they had lived by for ages, quietly loaded themselves with decades of unburned fuel, until they began to erupt into infernos too big to fight. We had a picture of nature at rest, and acting on the picture made the world more dangerous. The steady state we thought we were guarding had never been there to guard. Wherever we seek to find constancy, the ecologist Daniel Botkin wrote after a lifetime in the field, we discover change. There was never a still point that we alone disturbed. There was no garden. There was a storm, billions of years long, and we are the first thing that storm has ever thrown up that can look at the weather and set out to change it.

The point removes exactly one charge, the heaviest one: that we are the single blot on an otherwise perfect page. The page was never perfect. It was always being torn up and rewritten, and we are the first hand on it that can read.

There is a standard hidden inside the charge. To call a human act a desecration is to measure it against a pristine nature that our touch defiles—to treat the best state of the world as the one with the least of us in it, and every road and farm and city as so much subtraction from a perfect original. Alex Epstein has spent years promoting fossil fuels and energy abundance, and pressing the plain question that standard cannot survive: better or worse *for whom, and by what measure?* If the measure is how little we disturb a nature that was never at rest and never on our side—a nature that let the elephants strip Tsavo bare and would let us freeze and starve and drown with the same indifference—then the standard is not a defense of life at all. It ranks the barren valley above the terraced one and the malarial swamp above the drained field, because it counts only the absence of the human hand and never the flourishing of the human being.

Put the human being back at the center of the measure—judge our remaking of the world by whether it makes human life longer, safer, freer, and richer—and the whole picture turns over. The dam and the greenhouse and the vaccine stop being wounds in a sacred body and become what they always were: a naked, shivering animal doing the one thing it was born to do, which is to take a hostile world and make it fit to live in. The transformation is not the sin. It is the achievement, and the standard that calls it a sin has quietly chosen the empty world over the human one.

But putting the human being at the center of the measure is not the same as putting him alone in the frame. A drained swamp that ends malaria is a plain gain. An emptied fishery, an aerosolized chemical that thins the shield over every head, a species erased that might have carried the next medicine or simply the next reason to look up in wonder—these are costs by the human measure. The point was never that our remaking cannot wound us. It is that the wound and the healing are weighed on one scale, by what they do to the flourishing of the beings who must live with them, and the damage is answered the way this creature answers everything: not by laying down the tools and calling the transformation a sin, but by treating the harm as the next problem to solve. That measure counts the extinctions and the emptied seas as real losses. It still refuses to rank the barren valley above the terraced one.

Look back down the whole line and it is one act, worked on ever larger fates.

The cold, which the fox can only endure and be slowly shaped by, we dress against by nightfall. The sea, which drowns every other coastal animal that cannot outrun it, we wall out and then farm behind the wall. A plague, which culls whole species, we hunt to extinction and lock the survivor in a freezer. Each of these is a condition that every other living thing accepts as the terms of being alive—the given, the fated, the way of things—and that we, alone, pick up and turn over in our hands as a problem whose solution has not been found yet.

The Stoics taught that a life is measured by its quality and not its length, and that to rage against mortality is to rage against the terms of existence itself. Epicurus argued that death is nothing to the one who dies, since where death is, you are not. Every great faith has built an art of dying well, a way of meeting the end with grace instead of clawing. Bernard Williams made the case that an endless life would curdle into unbearable tedium, every possible experience eventually exhausted and repeated, meaning itself leaking away for want of an ending to press against.

And the nurses who sit with the dying know the sharpest version of all: that a medicine which treats death only as a defeat to be fought will fight it straight through a person's last good days, trading the peace of being home for one more grim week wired to a machine in a ward. The writers who have looked hardest at how we die now argue that our refusal to accept death has bred a new horror of its own—a way of ending that our great-grandparents, dying at home among their own people, would have thought barbarous. And the far pole is no lovelier: the man so consumed by the war on mortality that he cannot live, hoarding his days against an ending that sours every one of them in advance, is his own kind of warning. There is real wisdom in the acceptance of death, and real cruelty in its denial.

Take all of that at its full weight, and the turn still comes, because the counsel to accept has been given before—about every other fate on the list.

Accept the cold; it is the will of winter. Accept the plague; it is the will of God, or the way of nature, or the wage of sin. Accept the flood; the sea has always taken its share and always will. Every one of those acceptances was, in its day, the mature and pious and reasonable position, and every one of them was spoken over people who were already, quietly, refusing—weaving the coat, raising the dike, ringing the sick house with vaccinators. The refusers were called arrogant each time, and each time the arrogance turned out to be the truth arriving early.

The people who swore the famine could not be dodged were wrong. The people who swore the pox would always be with us were wrong. Death belongs to the same family as all of these, only larger and, so far, unbeaten—the last and hardest instance of the one problem this animal has been working since the first fire. Smallpox was a way of dying, and we ended it. Childbed fever was a way of dying, and cholera, and the infected cut, and the merely long winter, and we have pushed each of them back, one at a time. We did not accept those deaths. We solved the problems.

For nearly all human existence, a newborn could expect to live somewhere around thirty years; between a third and a half of all children died before they were grown, and a woman went into childbirth knowing it might be the thing that killed her. That was the ordinary condition of the species roughly unchanged from the first painted cave to the crowning of Victoria. Then, across about a hundred and fifty years—germ theory, clean water, the sewer, nutrition, anesthesia, the antibiotic—the ways of dying were forced back so far and so fast that the average human lifespan doubled, and the death of a child fell from the steady rhythm of every family into a rare and shattering exception. More than half of that gain is simply children who used to die and now grow up.

To call death a problem is not human arrogance reaching past its station. It is the plainest description of the animal that has been turning its fates into problems since the first night it refused to freeze.

None of that is a command to any one person to fight their own end past the point where fighting steals the peace of it; the grace of a good death is its own hard-won solving—and that grace is the truth of the single life, undiminished by anything that follows. What follows is the truth of the species.

We are the one piece of the billions-year storm that turns its fates into problems, and we have not yet made our peace with the largest fate, and every sign of our history says we were never going to. The acceptance is how a mortal creature lives well while the problem stays open; the refusal to call it final is how the problem gets worked at all. And the shadow that falls across the good hours—the sweetness of a thing loved because it will not keep—does not lift when the fate becomes a problem; it is the standing price of being the creature that can see its own ending coming, and it stays whether that ending is near or, one day, farther off than it has ever been. Death is the hardest of the fates. It is still a problem.

CHAPTER SIX

Solving Each Other

Put a married couple in a room, mention a subject they fight about, and leave them alone with a camera for fifteen minutes. From the tape, John Gottman can tell you with better than ninety percent accuracy whether they will still be together in six years. He is not measuring how much they fight—every couple fights. He is watching the seconds after it goes wrong: the moment one of them tries to bring the temperature down with a joke, a hand on the arm, an *I'm sorry, that came out wrong*—and whether the other one takes it.

That is the tell. Every couple ruptures. The ones who last repair.

Gottman ran that lab for decades, three thousand couples wired to heart monitors and filmed frame by frame, and the deeper he went the smaller the decisive thing got. It was never the grand gesture that held a marriage together—not the anniversary trip, not the renewal of vows. It was the tiny moment when one person makes a small reach for the other. Gottman called these bids, and every bid is answered one of three ways—you turn toward it, you turn away from it, or you turn against it. A wife glances up and says *look at that sunset.* Her husband can set down his phone and come to the window: he turns toward. He can grunt, still scrolling: he turns away. He can snap that he is busy: he turns against. It is the most trivial exchange imaginable, and it is the entire marriage in miniature, because some version of it will happen a hundred times today and tens of thousands of times this year, and the running tally of those answers is, more than any vow or any crisis, what the love will turn out to have been.

In the couples who were still happy six years on, partners had turned toward each other's bids most of the time. In the couples who divorced, they had turned toward barely a third. Love, for those first couples, was a thing they did—all day long, in ten-second increments, mostly without noticing they were doing it.

Most love dies the same way, in reverse. Not, usually, in a blaze of betrayal, though that happens; it dies in the slow accumulation of bids let fall—the small reaches, day after day, met with a grunt, a screen, a *not now*, until one of them stops reaching and neither can say afterward quite when the room went cold. Turning away is gentler than turning against, and far more common, and more lethal because it never looks like an attack. Couples rarely end because they fought too much. They end because, somewhere in the ten thousand small moments, they quietly stopped answering each other, and a bond that is not being built is already, silently, coming apart.

This is not how we are taught to think about it. We are taught that love is something you find. There is a person out there who fits, and the work, if there is any, is the searching—the swiping, the years of wrong ones—until the right one appears and clicks into place and the search is over. Every love story we are handed ends at the same instant: the two of them find each other, and the credits roll. The finding is the whole story. What comes after is assumed to run on its own, the way a found thing sits in your hand and stays found.

And the feeling that launched all those stories is real. Anyone who has been leveled by it knows it is no fiction—the sleeplessness, the whole world narrowed to a single face, the sense of having been struck by something from outside yourself. It is only not the thing that lasts, and it was never built to last.

Go back to the couples in Gottman's lab. The ones who lasted had not found some rare compatibility that spared them conflict. Roughly two-thirds of whatever a couple fights about never gets resolved at all—the same standing differences over money and sex and family and mess, argued in the same shapes, for fifty years. The lasting couples had simply learned to repair: to reach out in the middle of the fight, and, harder, to accept the reach when it came from the other side. The couples headed for divorce let the reach fall. One partner offered a hand, the other slapped it down, and the refusal taught the first to stop offering. What killed the bond, every single time, was the failed repair.

And Gottman found the thing that kills fastest: contempt. The curled lip, the eye-roll, the sentence shaped to say *you are beneath me*. Anger a marriage survives; contempt is the one poison that predicts the end almost by itself, because it does the exact reverse of a repair. It does not reach across the gap. It widens the gap on purpose.

Contempt is the first of what Gottman called the Four Horsemen; the other three are its cousins, and each has a repair that answers it. *Criticism*—the complaint that indicts the whole person instead of the single act, *you always, you never*—answers to a plain request for what you want. *Defensiveness*—the reflex that meets every grievance with a counter-grievance and admits nothing—answers to taking even a sliver of the responsibility on offer. *Stonewalling*—the wall of silence a flooded person throws up to survive the fight—answers to the quiet confession that you need a few minutes before you can go on. But contempt has no easy antidote, because contempt sits outside the argument altogether: it is a settled verdict about the other person's worth, and there is no repairing your way out of a verdict. The only thing that dissolves it is the slow rebuilding of the thing it killed—a practiced habit of noticing what is good in someone, kept up so steadily that the eye stops going first to what is beneath it.

Gottman also found why the repair is so hard to make in the moment, and the reason is in the body, not the character. When a conflict gets hot enough, the heart rate climbs past a hundred beats a minute, adrenaline floods the system, and the higher functions that would let a person listen and soften and reach go offline—the ancient machinery for facing a predator, switched on by a spouse across a kitchen table. Gottman called it *flooding*. A flooded person cannot repair, because the part of them that knows how has been temporarily shut down. The couples who did best were often the ones who had learned, without any theory, to stop and let the flood drain before they tried again. The reach requires a body calm enough to make it.

If love were found, intimacy would be a kind of discovery—you would meet the right person and recognize the closeness already waiting there. Yet intimacy can be manufactured, deliberately, between strangers, in about forty-five minutes, and Arthur Aron, a psychologist, built the machine that does it. Two people who have just met take turns asking each other thirty-six questions, in three sets, each more exposing than the last. It starts light—whom would you want at a dinner party—and climbs, question by question, into the material people usually guard: your worst memory, your mother, the last time you cried in front of another person. You answer honestly, then they do, back and forth, the vulnerability rising in step. At the end the two of them look at each other, saying nothing, for four minutes.

The order is the whole trick. Ask a stranger the last question first and they freeze; walk them up to it one rung at a time, each disclosure answered by one of equal weight, and they will follow you the whole way without alarm, because no step is ever taken alone. Vulnerability offered and vulnerability returned, kept in balance, is the ladder intimacy climbs—and it climbs the same way whether the two people are strangers in a laboratory or a married couple of twenty years trying, across a widened silence, to find each other again.

Strangers come out of it feeling close, sometimes alarmingly so. One early pair married; so did a writer who tried it on a date and caught the strange part exactly—that because the exposure rose gradually, she never noticed crossing into intimate country until she was already standing in it. The questions themselves are ordinary. What they expose is that closeness runs on a mechanism—reciprocal, escalating, answered disclosure—and that anyone willing to run it can generate the raw material of love from a standing start. Aron spent years correcting one misreading: the closeness does not come from the confessing. It comes from being heard—from the answer landing on the other side, the sense that the person across from you received what you handed over. The responsiveness is what builds the bond. Intimacy is something two people make between them, turn by turn.

Look at the arrangement from the far side of the world and the same truth shows through the other way. In much of history, and much of the world still, people did not marry the person they loved; they married the person they were given, and love was expected to arrive afterward, if it arrived at all—grown, over years, out of shared work and shared children and the daily grind of accommodation. The studies that have tried to compare the two are small and the findings are contested, so no one should lean too hard on them. But the pattern they keep half-finding is worth sitting with: in the love match, the feeling that blazed at the start tends to cool with the decades, while in the arranged match, where the couple began as near-strangers with a job to do, the feeling has sometimes been found to climb, until the two curves cross. And it points one way: the thing we treat as the precondition for the work may be the product of it.

❧

The deepest evidence that the bond is built and not found comes from the first bond any of us ever has, before we could speak or choose or go looking for anyone.

In the 1970s Edward Tronick filmed mothers playing with their infants and then asked the mothers to do one cruel thing: hold still. Go blank. Stop answering. Within seconds the baby—three or four months old—notices. It smiles harder, points, reaches, throws its whole small repertoire at that unmoving face, straining to start the dance again. When nothing works, it comes apart: twisting away, wailing, its heart and its stress chemistry climbing. Then the mother returns to life, and the two of them put it back together.

What Tronick saw on those tapes broke the picture of the flawless mother. Ordinary mothers and babies, he found, are out of sync most of the time—missing each other's cues, misreading, mismatching—and they pass their days in a constant low labor of repair, falling out of step and climbing back in. The security a child grows into is built from those thousands of small breaks mended, each one teaching it the single most useful thing there is to know about other people: that a rupture is survivable, that a connection lost can be found again. Tronick's conclusion was the one that cuts hardest: the child who is never failed learns nothing, because it never gets to practice the recovery. The mismatch and the mending work together—the small break, then the coming-back—and it is the pair of them that builds a person able to trust a bond at all. A child raised in perfect, unbroken attunement would be left, strangely, without the one skill that a whole life of love requires.

The pediatrician Donald Winnicott had reached for the same truth a generation earlier and called it the *good enough mother*—good enough, because she was not perfect, because the small failures and their mending were the thing the child needed to learn. And here is the detail that should stop anyone who imagines a bond as one person's gift to another: nearly half the repairs are made by the baby. The infant is already doing its share, turning back, catching the mother's eye, restarting the loop.

We are repairing our relationships before we can hold up our own heads.

That first pattern of rupture and repair does not stay in the nursery. Mary Ainsworth built a simple, devastating test for how it sets: *the Strange Situation*. Bring a mother and a one-year-old into an unfamiliar room, let the mother slip out for a few minutes, and watch what the child does when she comes back. Most protest the leaving and then greet the return with relief, settle quickly, and go back to playing—the pattern she called secure. Some greet the return with a cold shoulder, already schooled not to expect much from reaching. Some cannot be soothed at all, clinging and arching away in the same motion, having learned that the reach sometimes works and sometimes wounds and there is no telling which.

The three patterns are legible by a child's first birthday, and they track—loosely, but stubbornly—into the way that grown child will one day reach for a lover. The researchers who followed those one-year-olds into adulthood found that the shape of the earliest bond tends to lay down a template: a rough expectation, carried into every later love, about whether reaching out will be answered or will meet silence. The child whose bids were mostly met grows toward the confidence that closeness is safe; the child who learned that reaching brought either nothing or too much grows guarded, or grasping, in ways that outlast all memory of the cause.

It is a sobering finding, and it would be a bleak one, if the same research had not also found the escape hatch. The template can be rewritten. People raised in the coldest early weather turn out to be able to build security later, in a bond that keeps its promises long enough to overwrite the old expectation—earning, in adulthood and on purpose, the trust they were not handed as infants. Which is only the same lesson again, arriving from the far end of a life: the bond is made, and can be made even against a bad beginning, by the same slow work of reaching and being answered.

Which is where love is unlike every problem the reach has ever solved. The cold held still while we built against it. The sea stayed where it was long enough to be walled out. The plague did not revise itself to slip our nets faster than we could close them. Every fate we ever turned into a problem had the decency to hold still and be solved. A person does not hold still. The one you are straining to understand is, in the same moment, straining to understand you—remaking who they are as you remake your picture of them, so the target shifts every time you reach it, and the answer is never final and never yours alone. You solve, and you are solved back.

No other problem behaves this way. The mountain does not care that you are climbing it and does not rearrange its face to fool you on the way up. The person across the table has a model of you running behind their eyes, exactly as you have one of them, and each model keeps updating on the other, so that the instant you think you have them figured out they have already moved—grown, soured, forgiven, changed their mind in the night about a thing you both believed was settled. You are never finished knowing a person, because there is no finished person there to know; there is only a living process, revising itself, that you are invited to keep pace with for as long as the two of you last. That is the vertigo of love, and it is also its whole reward: the thing you are working to solve is the one thing in the universe that is working, just as hard, to solve you.

The poet Rilke, asked for counsel on love by a young man who wanted it to be simple, refused to make it simple. He called love the hardest task that is set for us, the work for which all other work is only preparation. He warned the young man off the easy version—the merger that feels like love and is really two people fleeing the labor of being a self by dissolving into each other. The good bond does the opposite, Rilke said: it asks each person to become more entirely themselves, not less, and to keep watch at the edge of the other's solitude rather than break into it. His picture of a strong love is two intact solitudes, each standing sentry at the border of the other—still whole, still separate, still finally unknowable, and choosing across that distance, repeatedly, to turn toward.

And the thing that gets built, given decades, is something the first blaze could never have been. Passion narrows the whole world to a single face and knows almost nothing about it; most of what it feels is its own heat. The bond two people make and break and remake across twenty years is a different order of thing altogether—unspectacular, and total. It is the condition of being completely known, every pettiness and cowardice and small-hours fear included, and chosen anyway, on the full evidence, by someone who has every reason to leave and stays. No stranger's ardor can come near it, because the stranger is in love with a guess. Only the long work delivers the thing this animal wants most and can find nowhere else: to be seen all the way to the bottom and kept.

This is why the work has no end, and why the endlessness is the point. Two people who keep choosing each other are two error-correctors turned on each other for life, each the other's endless, moving, half-solved problem, each revising toward a person revising in return. The labor is not the tax you pay to keep the love.

The work is the bond.

❧

If the work is the bond, someone will hear that *every* bond is worth the work—that leaving is failing at love, that a decent person just repairs harder, that the fault in a marriage grinding someone down must lie in their insufficient effort. That is false, and the same research that shows repair building love shows exactly where repair must stop. Contempt is a verdict. A bond running on it is already gone, whatever it looks like from the street, and the effort poured into it only sinks the person faster. The couples Gottman watched drown were pouring their strength into something that had already turned corrosive; the problem was never a shortage of effort. A high-conflict marriage does more measurable damage to a body than a divorce does.

The instruction is to repair what is reparable and to recognize what is not, and that recognition is its own hard skill, learned late if at all, and the lack of it keeps more people in the wrong rooms than anything else does. To say the bond is built is not to say every bond must be saved. It is to say you are always, in any bond worth keeping, answerable for your half of the building—and free, when the other half has turned to cruelty, to set the trowel down and go. No one owes a lifetime of repair to a person who has decided to wound them. Some of that work is the slow bricking-up of your own cage. Walking out of it is not a failure of love but an act of it, turned at last toward the one life you are responsible for.

When that difference is known, and the reparable bonds are kept and worked, the reward reaches the center of a life rather than its edges.

The needing runs deeper than feeling; it reaches into the wiring. James Coan slid married women into a brain scanner, warned them a small electric shock might be coming, and watched the regions that flare under threat. Then he let each woman hold her husband's hand. The threat response fell—the braced brain quieted, measurably, at the touch of a familiar hand—and it fell furthest for the women in the happiest marriages. A bonded human does not face danger alone; it offloads part of the load onto the person beside it and runs, in effect, on shared hardware. We use one another to regulate the machinery of staying alive, which is why losing a bond can feel like a physical amputation and in a real sense is one: a system that had been running on two processors thrown back, without warning, onto one.

It is the single largest thing that decides how a life goes. In 1938 a study began following a few hundred men—some Harvard students, some poor boys from the tenements of Boston—and it has tracked them, and then their wives and children, ever since: the longest look anyone has taken at whole human lives, cradle to grave, now running past eighty-five years. It measured everything a person might bet on—

money, fame, cholesterol, work, class, IQ. When the researchers finally asked what, at fifty, best predicted who would be alive and well and happy at eighty, the winner was none of the obvious things. It was the warmth of a person's relationships. Satisfaction with one's relationships at fifty predicted physical health at eighty more accurately than cholesterol did.

The researchers have said, more than once, that they did not believe their own numbers at first—that they kept waiting for the winner to be something a person could put in a bank or on a résumé, and it kept, stubbornly, coming back to love. One of the study's long-serving directors, near the end of his own life, was asked what eight decades of files on other men's lives had taught him. His answer came down to one thing: our relationships to other people are the only thing that finally matters. The men with good bonds lived longer, stayed sharper, and suffered less; the isolated ones sickened and died sooner.

None of this is only about marriage, and the study is careful on the point. The bonds that carried the men through were as often friendships and brothers and old comrades as they were spouses—the friend who knew you at twenty and still calls, the sibling you fought with for decades and never let go of. What mattered was never the label on the relationship. It was whether there was at least one person the man felt he could phone at three in the morning: one bond in which he was fully known and still held. A single tie like that can carry a whole life. Its absence is what does the damage, however crowded the room around it looks.

The size of that effect is easy to underrate. When researchers pooled the data from millions of people, they found that weak social connection raises the risk of early death about as much as smoking a pack of cigarettes a day, and more than obesity or a sedentary life. Loneliness is a mortal condition, and the cure is other people. The bond with them has to be built and tended like anything else that keeps a body alive—which is why the researchers took to calling it *social fitness*, a thing you are never done with, that goes slack the moment you stop working it, exactly like the muscles it is named for.

And we have built a world that lets the muscle waste. It is possible now to have every practical need met—food delivered, work done, a whole evening's entertainment, even a passable imitation of conversation—without once being answered by another person who knows you, and a rising share of people live that way: efficiently supplied and quietly starving. The ache they feel is no character flaw and no failure of gratitude. It is the body reporting a real deficiency, as concrete as scurvy, in something it cannot manufacture for itself and can draw only from other people.

❧

The animal that made a coat of a fox and a country of a sea meets, in another person, the one problem it can never finish and cannot bear to set down and gives a lifetime to a solution that keeps walking off and turning back. Nothing we do is harder.

And it does not stop at two. The same loop that ties one person to another—the reach across the gap, the rupture, the repair—is the loop out of which a third person is made. Two people who have learned to solve each other turn, together, to the longest and hardest problem either of them will ever take up: the raising of someone new, who arrives as unfinished as they once did, and who must be not merely kept alive but built into a person able to correct the world and, in time, to correct them. It is the highest-stakes work the loop will ever attempt, because a child is where the correcting animal is passed on—or stopped at the root, before it can defend itself. What a marriage risks is one bond. What a childhood risks is a mind.

CHAPTER SEVEN

The Unbroken Will

To form the minds of children, the first thing to be done is to conquer their will. A mother set that down in the spring of 1732, near the top of a calm and orderly letter written at her grown son's request, laying out the method by which she had raised her children. Susanna Wesley had borne nineteen and buried nine; the son who asked was John, who would found Methodism, and he thought her system worth keeping. Self-will was the root of all sin and misery; a parent who subdued it worked alongside God to save a child's soul, and a parent who indulged it did the devil's work and helped to damn it. Begin in infancy, and finish there: a will not broke early could not be broken at all.

She did not write a word of it in cruelty. She wrote it in love, as the considered counsel of one of the ablest mothers of her century, because she believed—as very nearly everyone who had ever raised a child believed—that a child arrives as a small unbroken animal, and that the central work of raising it is to break it.

Some version of it has run through most of the cultures that left instructions, in the proverb about the spared rod and the spoiled child, in the century of colonial schoolmasters who taught with a switch, in the twentieth-century experts who warned that picking up a crying baby would ruin it.

For most of human history the raising of a child and the subduing of a child were the same project, and the tenderest parents were the ones who did the subduing most thoroughly, on the sincere understanding that they were saving the child from itself.

The question worth asking is not whether Susanna Wesley loved her children. She plainly did. The question is what, exactly, the breaking breaks.

Because the will she set out to conquer was not a separate demon lodged in the child. It was the child's own engine—the same reaching, testing, guessing faculty we watched a baby run before it can sit up, the thing that says *goed* and asks why past the edge of the world. To conquer that will is not to remove a flaw. It is to teach the child the first and deepest lie there is: *that when your own judgment meets the word of authority, it is your judgment that must give way.* A child broken by that lesson learns to look for the answer not in the world but in the face of the adult holding the power—to read the parent for the verdict rather than reality for the fact. He is taught, before he can defend himself against the teaching, to distrust the one instrument that makes him human, and to hand its findings over to whoever outranks him. The conquered will is the first averted eye. It is the concrete setting in a mind too young to know what is hardening, poured by hands that love it and believe they are pouring a foundation.

And the lesson does not stay in the nursery. The child taught to hand its judgment to whoever holds the power grows into the adult who does the same—who looks to the authority instead of the evidence, who takes confidence for correctness and rank for reason, who has deferred so long that the machinery of independent judgment has gone stiff from disuse. Every closed mind and every closed society is built of adults like that, and most of them were made, in the first instance, exactly here—in a childhood where the price of one's own conclusions was set too high to pay. The conquered will is not only a private wound. It is the first and most dependable step in the manufacture of the deferential mind, which is the raw material every tyranny has ever run on.

The correction to all of this, when it finally arrived, came from an unexpected quarter—not from soft sentiment about children but from the hardest thinking about knowledge anyone was doing.

In the 1990s the physicist David Deutsch took his account of how knowledge grows to its conclusion in the one place almost no one had thought to take it, which was the nursery. With a collaborator, Sarah Fitz-Claridge, he set out a position they called *Taking Children Seriously*, and its starting point was as plain as it was radical. A child is a person. Its reasons and desires and its way of coming to know things work exactly as an adult's do, because there is only one way any mind builds knowledge—by guessing and correcting—and a small mind is running it as surely as a large one.

From which a single hard consequence follows. Coercion, Deutsch argued, jams the mechanism by which the child learns anything at all. To force a conclusion into a mind is to stop that mind from doing the one thing that turns a conclusion into knowledge: test it, doubt it, and hold the power to reject it. You can compel the behavior. You cannot compel the understanding, because understanding is what a mind builds for itself out of guesses it was free to throw away. The forced child does not learn the lesson. He learns to produce the sound of the lesson while the door to his own judgment quietly closes.

That is why the coerced conviction is hollow at the center, and why it so often falls the instant the pressure lifts. There was never a mind behind it, only a mind made to imitate one.

This lands on Susanna Wesley's project like a verdict. The whole aim of conquering the will was to install, by force, conclusions the child would then hold for life—about God, obedience, right conduct. And the deepest finding of the theory of knowledge is that this cannot be done. Force installs compliance, not conviction; the behavior, not the belief. What it reliably teaches, underneath the surface content, is the same lesson the schoolroom teaches in its own way: that the forward motion of your own mind is not to be trusted, that the reach must wait on permission. A generation of researchers has since watched the wreckage this leaves. The children most heavily coerced grow not into the most reliably good adults but, often, into the most brittle—outwardly compliant and inwardly ungoverned, obedient to whatever authority is present and lost when none is, because the internal instrument that was supposed to steer them was disabled young in the name of steering them. What the rod produces is not a moral compass. It is a child who checks the room.

There is an irony folded into the Wesley house that the method's admirers rarely sit with. The most famous product of that conquered will was John himself, the son who kept the letter—a man of monumental discipline and monumental unease, scrupulous to the edge of torment, drawn all his life toward authorities he could submit to and rules he could obey, and rarely at peace inside any of them. It would be too much to lay a complicated man at his mother's door; people are never that simple. But the pattern it points at is the one the record keeps turning up. The maximally governed child so often becomes the poorly governed adult—rigid where watched and lost where not, or in flat revolt against everything the rod stood for the moment the rod is out of reach—because the thing that was meant to be built inside him, the judgment that steers a person from within, was the one thing the whole method spent itself overriding.

Aaron Stupple's *The Sovereign Child* presses the point with a bracing purity: stop overriding the child's reasons with the raw fact of your power, and trust that the small person across from you is a person. What the theory of knowledge forbids is not the bedtime. It is the lesson underneath it. A bedtime can exist and leave the child's mind entirely free, if it comes with a reason and stays open to one. The question is whether the child is being taught that his mind is his own to steer by, or a thing that needs permission and can be overridden.

❧

The first half is to stop substituting your reasoning for the child's. The second half is to reckon with the quality of the reasoning the child will now do in its place. A mind left free but un-tooled is not thereby a good reasoner; it is only an unguided one. The task is to hand the child, patiently and across years, the tools of the corrector. How to tell a good explanation from a comfortable one. How to ask what would have to be true for a thing to be so, and what would show it false. How to notice when you are fooling yourself. How to hold a belief the way an honest scientist holds a theory, gripped hard and held open to the fact that could overturn it. These are not native in their finished form; the raw engine is native, but the discipline that aims it is learned, and learned mostly from other people, in the long apprenticeship of a childhood.

None of it is delivered in a lecture. It is handed over in the grain of ten thousand small exchanges. The child announces that the moon is following the car. The coercing parent, if it answers at all, corrects the fact—no, it only looks that way—and moves on, having taught nothing but that the parent knows and the child does not. The other parent asks the question that hands over a tool instead of a verdict: how could we find out? What would we see if the moon really were following us, and what would we see if it weren't? The fact hardly matters; the child will meet the moon again. What the child gets is the move, the turn from the claim to the test, and it arrives not by being taught but by being performed, in the open, until it is simply the way the child's own mind runs.

Put the two halves together and each rescues the other from its failure. Autonomy without the tools is the exposure the critics rightly fear—a child set free to be wrong forever with no method for finding out. The tools without autonomy are worse than useless, a set of instruments handed to someone forbidden to use them, correction preached by a parent who permits no correction of himself. Freedom is what makes the method matter; the method is what makes the freedom safe. Neither works alone, and the culture keeps offering them alone, which is why it swings between the tyrant and the doormat and calls each one, in its turn, good parenting.

The first commitment without the second is an abdication dressed as respect. The second without the first is a gift you will not let the child open.

The goal this points at is neither of the goals the old debate fought over. The authoritarian wanted an obedient child; the liberationist wanted a free one; and the target that matters is a third thing that contains what was true in both. The aim is a *self-corrector*—a mind you can trust with its own judgment because it has been given the means to check that judgment against the world. You are not raising a child who will do as he is told. You are not raising a child who will merely do as he pleases. You are raising the one kind of person who can be handed the gift of his own freedom without being destroyed by it: *the kind of child who knows how to find out when he is wrong and correct it.*

What builds a secure child is not a parent who never fails. It is a parent who fails and repairs. The child learns, in its body and before it has a word for any of it, that a rupture is not the end. The parent who is never wrong, or who can never be seen to be wrong, does not give the child a model of perfection to aim at. He gives the child a model of the sealed mind—authority that does not answer to reality, certainty that admits no correction—which is the exact template of every closed and dangerous mind there is, installed at the root, in the person the child trusts most.

Which is why the single most powerful thing a parent can do is also the thing the tradition of conquering forbade absolutely. It is to say, to the child, and mean it: *I was wrong. I was wrong, and you were right, and I have changed my mind.* Susanna Wesley could no more have said it than flown; the whole architecture of the conquered will depends on the parent's judgment standing beyond appeal. But a parent who can say it, and be seen to say it, hands the child the most valuable demonstration it will ever receive—that even the largest authority in its world holds its beliefs open to the facts, that changing your mind in the face of the evidence is not the weakness of the ruled but the strength of the free, that the door stays open even at the top. The child does not learn correction from being corrected. It learns correction from watching the people it loves correct themselves.

The words cost the parent something real, which is the whole reason they teach so much. A child can tell the apology that manages him from the admission that cost something. A parent going through a motion looks nothing like a parent visibly changing his mind because the facts made him. The second is rare, and a child who receives it even a few times keeps the memory of it for life, because it proved to him, once, from the person best placed to prove it false, that the door at the top of his world was open after all.

That is the second womb finishing its work. The first womb, of flesh, hands over an unfinished body. The second, made of other people, finishes the mind—and it finishes it not by pouring in the right conclusions but by modeling, ten thousand times in the ordinary traffic of a home, the one motion that will have to serve the child for the rest of its life in every problem it meets and every person it loves: the reach, the error, the honest turn to the fact, the mending. A family is the first place a human being sees that motion performed or does not. It is where the corrector is made, or broken, before the child has any say in which.

It is here, too, that the long inheritance is handed on—the piled-up knowledge of everyone who came before, which no child could reason back into being from nothing and so must take, on trust, from the adults around it. But there are two ways to hand it down, and they make two different creatures. Hand it down by command, as a set of conclusions to be swallowed and not questioned, and you make a mind that holds the inheritance the way a servant holds an order: intact, perhaps, but sealed, unexamined, unable to add to it or repair it. Hand it down the other way, as the best guesses of fallible people who are still busy correcting them, and you make a mind that receives the whole inheritance and holds it open, fit one day to carry it further. The same knowledge passes down either way. Only one way passes down, with it, the power to improve it.

❧

To raise a child well is to pour years of your one life into a person who will walk out of it—to build, deliberately and against every impulse toward control, something that is meant to outlast and outgrow and finally leave you, and to find the meaning of the labor in its being for a future you will not fully share. It is a reach past the edge of one's own life, a love aimed down the length of time at someone who cannot yet thank you and may never know what it cost.

No other animal does this. A creature that raises its young at all raises them into copies of itself, fitted to the world it already knows; only the human sets out to raise a child equipped to outthink him—to reach conclusions he never reached, correct errors he never caught, and render the parent, in the end, unnecessary. To raise a corrector is to build, on purpose, a mind that may one day overturn your own, and to want it to.

The reach a family teaches first does not stop at the edge of a single life. The same power that shapes a child for a future beyond the parent raises cathedrals for a future beyond everyone—things no builder will ever see complete.

CHAPTER EIGHT

The Cathedral and the Rocket

Six hundred and thirty-two years. That is how long it took to build the cathedral at Cologne—and a stonemason who cut a block of limestone for it in the year 1300 knew, as he cut, that he would never see it finished. His father had worked the same site, and his son would after him, and none of the three would live to watch the towers rise. The plan they cut stone from had been drawn by a man already fifty years dead. The plan they cut stone toward would not be fulfilled for another five hundred and eighty years.

The foundation stone went down in 1248, to shelter the bones of the Three Kings and to out-soar every church in Christendom. Work crawled, and then, around 1560, it stopped. For three centuries a medieval crane stood frozen atop the half-built south tower, a wooden skeleton against the Rhine sky, and the people of Cologne grew so used to it that a saying took hold: when the cathedral is finished, the world will end. In their minds it was never quite meant to be done. Then in the 1840s a unifying Germany went looking for a national symbol, pulled the surviving medieval drawings out of storage, and finished the thing exactly as its first architect had imagined it—in 1880, six hundred and thirty-two years after the first stone, briefly the tallest structure on earth.

When the bombers came in the Second World War and beat the city around it into a field of rubble, the cathedral took fourteen direct hits and did not fall; photographs from 1945 show it standing gaunt and scorched and whole above the wasteland, the one vertical thing left, as if seven centuries of gathered intention had made it harder to kill than the modern city at its feet. It has hardly left scaffolding since. Wind and pollution eat the stone as fast as masons can cut its replacement, so the work that began in 1248 has, in a real sense, never once stopped.

Cologne was one spire in a forest of them. Across three centuries the towns of medieval Europe threw themselves into a building spree with almost no parallel in history—by one reckoning, the masons of France alone quarried and cut more stone in those three hundred years than the ancient Egyptians moved in the entire age of the pyramids. A single cathedral could swallow the surplus wealth of a whole town for a hundred years, and a good many were begun in confidence and left as stumps when the money or the faith ran dry. A few went up fast: after fire gutted Chartres in 1194, the townspeople rebuilt it in a startling twenty-six years, dragging the stone up from the quarries on their own shoulders in what the chroniclers called a cult of the carts. But the fast ones were the exception. The ordinary unit of these projects was the lifetime, and rarely a single one—a chain of lifetimes, each worker handing the unfinished thing on to a stranger he would never meet.

A cathedral like that was the largest coordinated undertaking its world knew how to attempt, and nothing about it was simple. The man in charge was the master mason, and he was no mere bricklayer; he held the whole design in his head and on parchment, commanded the lodge of carvers and glaziers and carpenters, and drew plans he handed down to a successor not yet born, the way a dying general hands on a campaign. The Gothic style those men invented was itself a feat of distributed problem-solving—the flying buttress, that spidery external strut, carried the crushing outward thrust of the stone roof away from the walls and down into the ground, so that the walls no longer had to hold the weight and could dissolve, for the first time in history, into great sheets of colored glass. It let them build light.

And it let them overreach: at Beauvais, the masons pushed the vault higher than anyone ever had, to a hundred and fifty-seven feet of impossible stone, and in 1284 it came crashing down, and the cathedral was never finished. The reach at scale can grasp past what it can hold. That failure belongs in the story beside the triumph, because they come from the same nerve.

How the medieval builders, most of them illiterate, coordinated a project longer than a life is itself a marvel. The knowledge lived in the lodge—the workshop beside the rising walls—and in the bodies of the men, passed hand to hand as apprentices became journeymen and journeymen became masters, the geometry carried in templates and habits rather than in books. Master masons were the celebrity engineers of their age, summoned across a continent, and the tricks they carried traveled with them, which is why a technique worked out at one cathedral would surface, a decade later, three hundred miles away. The design outlived the designer because it was never wholly in his head to begin with; it was distributed across a craft, a tradition, a set of trained hands that could lose any one member and carry on. A cathedral was a mind larger and longer-lived than any of the minds that made it.

Generation after generation poured their entire working lives into something none of them would ever use, for the benefit of strangers not yet born—people who would never learn their names, never thank them, never exist in any world the builders could touch—and the hard-nosed account says, flatly, that people do not do this. We act for ourselves and our own, for rewards we can collect, on horizons we can see. The argument has real teeth, and it deserves them. Every instinct we carry was shaped by whether it helped its owner survive and breed, and no gene ever profited from a lifetime of labor poured out on people six centuries downstream who share nothing with the builder and can repay nothing at all. A creature built by natural selection ought to be, at the root, a machine for its own advantage—and the coldest readers of human nature have always insisted that this is what we remain, that the monument is vanity, the sacrifice a bid for status among the living, and the whole grand language of legacy and posterity a costume thrown over motives no different from any animal's.

Grand talk about building for the ages is the kind of thing carved onto a monument after the fact; the real motive is always nearer and smaller. And the great works, when you look closely, come from great individuals anyway—one architect, one genius, one name on the plaque.

Both halves of that are wrong, and a cathedral is only the first place you can watch them fail.

Six centuries after Cologne's foundation stone, a different kind of cathedral went up in eight years. When John Kennedy stood at Rice University in 1962 and bound his country to the Moon, he was honest about what he was promising: a rocket more than three hundred feet tall, taller than the football field he spoke on, built of metal alloys some of which had not yet been invented, fitted together with a precision finer than the best watch. The means to keep the promise did not exist when the promise was made. He named it for what it was—an act of faith and vision, he said, for we do not now know what benefits await us. We choose to go to the Moon, he told the crowd, and do the other things, not because they are easy, but because they are hard.

He was speaking into fear at least as much as into hope. The Soviet Union had already put the first satellite and then the first man above everyone's heads, and a rattled America wanted the high ground of space badly enough to spend, at the program's height, nearly a nickel of every federal dollar chasing it. But fear can commission a program; it cannot build one.

Keeping that promise took four hundred thousand people. Not four hundred thousand in sequence, but four hundred thousand at once, across twenty thousand companies and universities: machinists, mathematicians, metallurgists inventing the alloys Kennedy had promised out of thin air, technicians, test pilots, and a programmer named Margaret Hamilton running the team that wrote the code the whole descent would hang on.

On the twentieth of July 1969, it worked.

Not one of those four hundred thousand people could have told you how the whole machine worked. The woman stitching the spacesuit did not know the guidance equations; the man who wrote the guidance equations could not have cast the engine bells; the metallurgist who poured the alloy had never seen the mission entire. The Saturn V had something like three million parts, and no single skull on earth held a working picture of all of them at once. A machine that no one person understood carried three people to another world and brought them home.

It existed only as the sum of four hundred thousand partial understandings, held together by schedules and specifications and trust, each person minding their corner and believing the others were minding theirs. This is how anything past a certain size gets built. The lone genius is a story we keep because a single face fits on a stamp and a collaboration does not.

Look closely at any fragment of it and you find the same strangeness. The computer that flew the astronauts to the surface was feebler than a modern doorbell, and its program was a thing you could hold in your hand—physically woven, wire by wire, by women in a Massachusetts factory threading copper through or around tiny magnetic rings, a one for through and a zero for around, until the software existed as a hand-knitted textile. The spacesuit that walked on the Moon was sewn by seamstresses at a division of the company that made girdles and bras, twenty-one layers stitched by hand to a sixty-fourth of an inch with no pins allowed, because a single stray needle left in the fabric could have killed the man inside it.

And when the lunar module was falling toward the surface with minutes of fuel left, the computer suddenly cried alarm—overloaded, screaming that it could not do everything being asked of it at once—and it did not crash, because Hamilton's team had built it to triage under overload, to shed the trivial tasks and guard the ones that were keeping men alive. The landing happened because a programmer had thought, years earlier, about what her machine should abandon in the instant it began to drown.

There is a photograph of her from those years: a young woman in a print dress standing beside a stack of the program listings her team had written, and the tower of paper rises higher than her head. That stack flew to the Moon. So did the hand-woven memory and the hand-stitched suit. The most futuristic thing the species had ever attempted was, at the level of the actual work, astonishingly handmade—thousands of people doing careful, specific, mortal labor, not one of them able to see the whole, all of them somehow pulling the same direction.

There is an old parable about a traveler who asks three masons what they are doing: the first says he is cutting a stone, the second says he is earning a wage, and the third says he is building a cathedral. It has a Space Age twin, in which a janitor at NASA, asked by the visiting president what he does, answers that he is helping put a man on the Moon. Both stories are almost certainly buffed up in the telling; the janitor exchange has no record, and the bricklayer tale is a chestnut. What they point at is documented to the page. A management scholar went through eighteen thousand pages of NASA's internal records and traced exactly how a mission gave hundreds of thousands of ordinary workers, most of them doing narrow and unglamorous jobs, a single shared picture of where it was all going—so that a man with a mop could say, and mean, that he was putting a man on the Moon, because he was. The parable is a fiction. The four hundred thousand are not.

Beneath the border of France and Switzerland runs a ring seventeen miles around, the largest single machine our species has ever built, chilled to a shade above absolute zero, where something like ten thousand physicists from a hundred nations—most of whom have never met—fire beams of protons into one another trying to discover what the universe is made of. When they finally caught the particle, they had spent thirty years and ten billion dollars hunting, no single person had built the detector, grasped the whole apparatus, or could stand up and claim the discovery. It belonged to the collaboration: a cathedral of the mind, raised by a congregation far too large to kneel in any one nave. The thing being reached for had changed, over the centuries, from God to the Moon to the first flicker of creation. The reaching had not changed at all.

To be one hand among many on a thing larger than yourself is the surest source of meaning our kind has ever found—steadier, for most people, than any private success. The janitor who believed he was helping reach the Moon saw his own small task truly, in its real setting, which is more than most people manage on most days. A life spent entirely on itself stays small however much it piles up. A life folded into something that outruns it takes on that thing's size. This is why such work feels like a privilege to the ones who do it, and why its opposite, labor that serves nothing past the wage, quietly corrodes the person performing it.

It would be easy to file all of this under the past—the cathedral as medieval religious mania, Apollo as Cold War chest-beating—and conclude that people no longer build for a future they will not live to see.

It is true that we have built a civilization that runs on quarterly earnings and the election cycle, that discounts the future so steeply a decade out can look like never, and a steady chorus insists the old capacity has gone with the cathedrals—that no age this distracted, this short of breath, could lay a foundation stone for the year 2600 and mean it. The chorus is wrong, and the refutation is a handful of things being built, right now, by people who have made their peace with not living to see them done.

Start with the loudest of them. A company called SpaceX has spent more than two decades, and a fortune, building a rocket for a destination the people building it do not expect to see settled in their lifetimes. The rocket is called *Starship*, the largest and most powerful ever flown, more than four hundred feet from base to tip, and its whole design turns on one stubborn idea: that a rocket, like an aircraft, should be flown and landed and flown again, instead of thrown away after a single use as every rocket before it had been. The company has already done the part that looked impossible, bringing a booster the height of a skyscraper back down out of the sky and catching it in the air between a pair of mechanical arms called chopsticks. It fails often and in public. Its rockets have burst into fireballs on live video; the program runs years behind its own promises. None of that is hidden, and none of it stops the work, because failing in the open and flying again is not a setback in the method. It *is* the method—the same loop a baby runs with a dropped spoon, scaled to the size of a moon shot and far beyond.

But the rocket is not the cathedral. The rocket is the scaffolding. What it is being raised toward is a thing its builders state plainly and without any expectation of seeing finished: a city on Mars, self-sustaining, large enough to carry on if the ships from Earth should ever stop coming. Elon Musk's plan is measured not in fiscal quarters but in centuries, a foundation stone set for a completion no one now alive will witness. And the stated reason is plain. Life has only ever had the one address. Everything that has ever lived has lived in a single place, and a single place can end—by a rock from the sky, by our own hand, by any of the accidents that deep time turns from unlikely into certain. To set a living, breeding, self-repairing branch of humanity somewhere else is to make a backup of the one thing we have ever found that can make a backup of anything, which is a mind. Whether it comes on the promised schedule hardly matters; the Mars plans have already slipped, pushed back to clear the way for a return to the Moon first. Thousands of able people, and the owners of hundreds of billions in capital, believe in it as firmly as the cathedral builders believed in theirs.

❧

Under the permafrost of a Norwegian island eight hundred miles from the North Pole sits a vault holding more than a million seed samples: a backup copy of the world's food crops, sealed inside a mountain against a catastrophe its builders pray will never come, engineered to keep its charges alive for a thousand years. Every nation on earth, whatever it makes of every other, keeps a copy of its agricultural birthright in that one frozen room, on a shared understanding that the harvests of the deep future weigh more than the quarrels of the present—an act of trust among peoples who agree on almost nothing else. It has been drawn on in earnest for one emergency only, to send seeds back to war-wrecked Syria so that a bombed agricultural bank could be rebuilt from the frozen duplicates.

In a mountain in west Texas, meanwhile, engineers are assembling a clock designed to keep time for ten thousand years, ticking once a year, chiming once a century, sounding a note no living person will be there to hear, for an audience composed entirely of the unborn. Its designers state the purpose without a trace of embarrassment: they hope that whoever stumbles on it, thousands of years from now, will understand that we built it because we cared about them.

❧

Bury nuclear waste that will stay deadly for ten thousand years, and you face a problem no engineer was ever trained for: how to warn a person who will not share your language, your alphabet, or any symbol you now possess—across a gulf of time longer than all of recorded history—to keep away from this ground. Panels of linguists and anthropologists and artists have argued it in dead earnest: whether to raise a field of menacing concrete thorns, whether a skull will still read as danger or come to read as treasure, whether to found something like a priesthood to hand the warning down the generations as ritual after every language it was first written in has died. They are composing a message to descendants a hundred centuries off, people they cannot picture, for no reason but to keep them from harm. It is the cathedral impulse turned to its grimmest use—love for the unborn, carved into the rock against deep time.

Something in us reaches clean past the horizon of our own existence and fastens, in earnest, onto people who do not yet exist. There is a name for a force that spends itself on the good of someone who can never repay it. Aimed at the person beside you, we call it love. Aimed down the length of time at strangers, it is the same thing, and it is as real a feature of the animal as hunger—only rarer, and finer, and ours alone.

Which is where the cathedral and the rocket stop being opposites. We are trained to hold them apart—the cathedral as faith and the rocket as science, one raised toward God and the other toward vacuum, one built on prayer and the other on physics. Set them side by side and the difference thins to nothing. Each is a single creature building a thing larger than any one life, across more time than any one life holds, toward a completion no builder is promised to see. Each runs on a faith the hard-nosed account swears we do not possess: that the future is worth the present, that people we will never meet are worth our labor now.

The engineer at Cape Canaveral may have been baffled to hear that he shared anything with a thirteenth-century stonecutter, and the stonecutter would not have understood one word the engineer said. Hand each of them his tool, though—the chisel, the wrench—and point him at a thing built to outlast him, and the two are performing the identical human act, the act nothing else alive performs: spending a life on a future that belongs to someone else.

The cathedral and the rocket are the same act, pointed at two different heavens.

Only the address changed. For most of our history we aimed the impulse straight up, at the god above the roof of the world. Then the roof came off, and we aimed the identical impulse at what lay behind it—the actual heavens. Antoni Gaudí, asked why the vast church he had given his life to would not be finished until generations after his own death, was untroubled by the arithmetic. My client, he said, is not in a hurry.

The line is a joke and an entire theology at once. Gaudí knew he was one mason on a work measured in centuries, and he found in that not despair but a strange kind of freedom—released from the tyranny of finishing, free to labor for a horizon so distant it turned his own death into a detail. That freedom is on offer to anyone who takes up a piece of a thing larger than a single life, and it is one of the few real answers our kind has ever found to the shortness of its years.

No other animal has ever laid a stone for a wall it knew it would never see standing. That plain fact—that we work, routinely and gladly, for a finish line beyond our own graves—is among the strangest and largest things about us, and it goes unnoticed only because it is everywhere. We live inside the cooled ambition of the dead: their roads, their laws, their books, their medical discoveries, their cathedrals, the words in this sentence, all of it raised by people who knew they were building for us and did it anyway. To be human is to be born into a house you did not build and to add a room you may not live to use.

No other creature has any relationship at all to the unborn or the long dead. A wolf inherits nothing from its grandparents but its body and leaves its great-grandchildren nothing but the same; every generation of every other animal starts the world over from zero, in the same unchanging present, forever. We alone live inside a chain of hands reaching backward and forward out of sight—taking, unearned, the roads and words and rights and remedies the dead worked to leave us, and setting our own stone into a wall whose top course we will never see. That chain is what we mean by civilization, and it is held together by nothing sturdier than the willingness of each link to labor for the next.

The capacity did not come free, and nothing else on earth stumbled into it, because it demands a stack of improbable things all at once. You must be able to imagine a world you will never occupy, and then to care what happens in it. You must trust strangers you will never meet to carry on a job you began and cannot finish. You must be able to hold a plan in common with the unborn, in symbols durable enough to outlast the hands that drew them. Take any one of these away and the cathedral cannot rise and the vault cannot be sealed. That we possess all of them together and use them so casually that we have to be reminded they are remarkable, is the quiet miracle underneath every monument ever raised.

And yet the cathedral, for all its grandeur, is only the most visible shape the reach past the self can take—strangers bound to one labor across the reach of *time*. There is a quieter binding, one that raises no spire and is never finished and leaves nothing a person can stand inside. It ties far more strangers together across the reach of *distance* than any faith or any tongue ever gathered, and it is at work, unnoticed and unthanked, in the very city where the cathedral stands, on any ordinary morning of the world. It is also the one whose breaking can starve that city without a single soldier at the gate.

CHAPTER NINE

The Cooperation of Strangers

On any ordinary morning, a great city eats. Ten million people in London or New York wake and are fed—bread, coffee, eggs, oranges picked a thousand miles away—and almost none of them spent a moment the night before wondering whether the food would be there. It is simply there, in something close to the right amounts, in ten thousand shops, having crossed oceans and mountains from farms and boats and warehouses not one of the eaters will ever see. No one oversees this. There is no Ministry of Breakfast, no official whose task is to see the city fed, no master plan filed in an office somewhere. Go looking for the person responsible for feeding New York and you will not find them, because they do not exist. And yet the city is fed, every single day, more reliably than almost anything any government does run.

This is so familiar that it takes an effort to see how strange it is. Consider something smaller: a plain wooden pencil, the cheapest object in the shop. No one on earth knows how to make one. The claim is not original—an essay from the 1950s by Leonard E. Read made it by following a single pencil back through its makers—and it sounds absurd until you follow the pencil back.

The wood is cedar, cut in the Pacific Northwest, felled with saws made of steel smelted from ore mined somewhere else, by men who ate food grown by strangers and drank coffee from Brazil. The graphite came from a mine in Sri Lanka; the yellow paint, the glue, the tiny brass ferrule, the rubber tip made from latex tapped in Malaysia—each of these runs back through its own endless tree of hands, and no single mind in that whole vast collaboration knows more than a sliver of it. The man who runs the graphite mine cannot fell a cedar or mix the lacquer. The chemist who formulates the eraser has never seen the forest. And still, from the coordinated labor of hundreds of thousands of people who do not know one another exist, who speak different languages and serve different gods and would not agree on a single political question, a pencil appears on a shelf, and it costs almost nothing, and you do not even have to say thank you.

This is *the division of labor*. No one makes a pencil, or a meal, or a life, alone; each of us does one small thing and trades for the rest, and civilization rests on the plain, world-altering discovery that a hundred people each doing one task and swapping the results will produce beyond all comparison more than a hundred people each trying to do everything.

Ludwig von Mises pressed this further than anyone, to a conclusion that sounds strange only until you sit with it: that society itself, at bottom, is the division of labor. What binds human beings into a shared life is not, primarily, blood or love or a common god or a common ruler—it is the discovery that we are each made richer by the others' work, that a stranger's specialized labor is worth more to you than your own hand at the same task. It is a bond that asks the two parties neither to know nor to like nor to agree with each other about a single thing. It asks only that each have something the other wants. And that is enough to knit a tapper, a logger, and a child at a shop counter into one cooperating whole—a stranger and sturdier union than affection ever built, because it does not depend on affection, and survives the plain fact that almost everyone inside it will never feel a flicker of goodwill toward anyone else in it.

The mechanism is one of the largest and least noticed instances of the correcting animal at work—the same guessing, testing, and correcting a child runs in a crib, working now across a whole civilization at once, coordinating more people toward more ends than any mind could hold, and doing it without a mind in charge at all.

One of the deepest things anyone has ever said about economics was said in Vienna in 1920, and it was not a claim about incentives or human nature. It was a claim about knowledge. Mises announced that a society that abolishes markets in the means of production does not thereby get a fairer economy or a slower one. It gets a blind one. Without real prices—prices set by real people spending their own resources—there is no way to calculate whether one use of a ton of steel is wiser than another, because there is no common measure of the countless things the steel might become and the countless people who might want them. The planner who wants to do good, who has seized the factories to run them for everyone's benefit, discovers that he has burned out his own eyes. He can issue orders, and the orders will be obeyed, and he will have no way whatever of knowing whether they were the right orders, because he has destroyed the only instrument that could have told him. A planned economy, Mises said, cannot do arithmetic.

In the Vienna of 1920 this was heresy. Socialism was the future, as every advanced mind understood, and the economist who announced that the future could not count was treated the way sealed institutions always treat their correctors. Among the young men certain that Mises must be wrong was a decorated veteran named Friedrich Hayek, who had gone to the Italian front at eighteen, spotting artillery fire from an open airplane, and come home half-deaf in one ear, mildly socialist, and set on the question of how a civilization holds itself together or comes apart. He came to Mises's private seminar expecting to have his socialism confirmed, and had it taken away instead. Then he spent the next twenty-five years discovering that his teacher had understated the case.

Mises had shown that a planner given every fact would still be unable to compute with them. Hayek saw that the planner never gets the facts. Everything a society would need to know in order to plan its economy—who wants what and how badly, what is scarce this week and abundant the next, which mine is flooding and which harvest failed, what each of a billion people knows about the particular circumstances of their own time and place—all of that knowledge exists, but it does not exist anywhere in one piece. It is scattered across every separate mind, most of it unspoken, much of it the kind of thing a person could not put into words if asked: the feel a good machinist has for when a bearing is about to go, the hunch a trader plays, the thousand small preferences no one has ever articulated even to themselves.

No planner, however brilliant, however well equipped with computers, can gather that knowledge into a central office, because most of it cannot be written down and all of it is changing faster than it could be collected. The two arguments are one verdict delivered twice: the knowledge cannot reach the center, and if it could, the center could do no arithmetic with it. The central planner is not merely overworked. He is, in the most exact sense, in the dark.

The market solves this impossible problem, and it solves it with prices. A price is a single number into which the dispersed, unspoken, uncollectable knowledge of millions has been boiled down. When tin grows scarce somewhere in the world—a mine floods, or a new use is found for it—the people at the flood and the people who found the new use do not have to announce anything, or understand one another, or even know that the other exists. The price of tin rises. And that one rising number travels out through the whole system, and a hundred thousand people who have never heard of the flood, who could not find the mine on a map, quietly begin to use a little less tin, to look for substitutes, to ration what they have—each of them responding not to the event, which they know nothing about, but to the number, which tells them everything they need to do without telling them why. The price is a signal that carries knowledge no one possesses in full. It is a telegram from the whole of reality to each small part of it, and it lets a butcher in one country act correctly on a fact he will never learn, occurring in another country he will never visit.

Thomas Sowell has spent half a century pressing this insight into every corner of social life, and he came by it the hard way: raised by a great-aunt in Harlem, a high-school dropout who carried a camera through the Korean War as a Marine, he arrived at economics a convinced Marxist and was talked out of it not by an argument but by a summer of Labor Department data showing what a minimum wage did to the very workers it was meant to help. His conversion is the argument compressed into one life—a man held a theory, reality returned a correction, and he took it.

And the edge he added to Hayek is a question of geography. What decides whether a society thrives, Sowell argued, is not how much knowledge exists inside it—every society holds more than any one mind could ever use—but where the knowledge sits, where the decisions get made, and whether those are the same place. A market keeps them together: each decision stays in the hands of the person holding the specific, local, half-wordless knowledge that decision needs—the shopkeeper who knows his own street, the farmer who knows his own soil, the buyer who alone knows what the thing is worth to him this afternoon. Central planning pulls them apart, gathering the decisions into a room the knowledge can never reach.

A price, then, is more than a signal. A price is a correction. It is reality pushing back—the world's way of telling a mind that has guessed wrong about what things are worth that it had better think again. Raise the price of a thing you are making above what anyone will pay, and the loss corrects you; find something people need that no one is yet supplying, and the profit rewards the guess and pulls others to copy it. Profit and loss are the feedback of an error-correcting system—the market's version of the fact that stops the child's wrong theory cold.

An entrepreneur is a person making a conjecture, in public, with his own resources, about what strangers will want and what it will cost to give it to them; and the market is the pitiless, indispensable test of that conjecture against reality. Most of the guesses are wrong. The system works not because the guesses are good but because the wrong ones are caught and killed and the right ones are copied and spread—which is, exactly, the only method by which knowledge of any kind has ever grown. The market is the conjecture-and-correction loop with a civilization inside it.

This is the machinery Mises said the planners were smashing. He predicted, while the whole world called him a reactionary crank, that an economy stripped of prices would lurch from surplus to famine, building steel mills no one needed beside shelves no one could fill—and the whole bloody twentieth century was, among its other horrors, the running of that experiment, again and again, to the result he named.

The man who said it was made to pay for the saying. Mises never received the professorship his work had earned; he taught in Vienna for decades without a salaried chair. When the Nazis took Austria, the Gestapo raided his apartment and carted off his papers, which vanished—until 1991, when they surfaced in a formerly secret archive in Moscow. The state built on planning had kept, filed and guarded in a sealed room for half a century, the papers of the man who proved that planned states are blind. He finished his career as an unpaid visiting professor in New York, unfashionable and unbending, and died in 1973, one year before the Nobel Prize went to his student Hayek for the line of argument the old man had opened five decades earlier.

And calculation was not even his central idea. In the treatise he called *Human Action*, Mises rebuilt economics from a single observed fact: that a human being acts. He feels the present as not good enough—Mises's term was *felt uneasiness*—pictures a state better than the one he stands in and reaches for means to close the gap. Action begins in dissatisfaction and proceeds by conjecture. Every purchase is a guess about what will satisfy; every venture is a guess about what strangers will want; and because the future is closed to certainty, Mises held that every acting person is an entrepreneur, staking something real on a picture of tomorrow that tomorrow will confirm or wreck. Profit is the confirmation and loss is the wreckage, and neither is a moral accident. The driest of the sciences, rebuilt from its foundations, turned out to rest on the same creature the crib revealed: a being that lives by imagining the world better than it is, acting on the image, and letting reality mark the exam.

So, a market is a correcting system, and central planning is the refusal of correction raised to the scale of a state. It fails for the identical reason the sealed mind fails, and the tyranny fails, and the ideologue fails: it has made itself unable to find out that it is wrong. The famine at the end of it is neither accident nor sabotage. It is what blindness looks like when it oversees a harvest.

There is a second thing the market is, past its genius as a machine for knowing, and Ayn Rand saw it more clearly than anyone: it is a *moral* arrangement. A trade is the one human transaction in which both sides win by consent. When two people exchange freely—this loaf for that coin, a day's labor for a day's wage—each is betting that they are better off with what they are getting than with what they are giving up, and both can be right, because value is not a fixed lump to be divided but a judgment each mind makes for itself. No one is looted. No one is commanded. The only alternative to the trade is the taking, and the taking runs on force.

Rand called this *the trader principle*, and she refused to rest capitalism's case on the ground everyone else chose—that it works, that it outproduces its rivals, that it feeds the city. It does all three, and she knew why. But for her the production was the corollary. The system's title to exist is that it is the only one built on the barring of force, and the barring of force, she held, is the precondition of civilization itself—because a mind cannot be corrected at gunpoint. It can only be silenced.

Which means the two halves of the case are one argument. Hayek and Mises proved that the market sees because no one commands it. Rand saw that the not-commanding is the moral point. A society that bars force has agreed to deal in reasons—to treat every person as a mind to be persuaded rather than an obstacle to be moved—and dealing in reasons is the only condition under which error can be found and fixed. The market knows for the same reason it is just: because every mind inside it is free to say no.

Mises drew a consequence from the trader's arithmetic that reaches past economics entirely. If a stranger's labor and land and ingenuity enrich you most when he is free and alive and prospering, then the ancient logic of conquest begins to dissolve. For most of history the foreigner was prey—his goods could be seized, his fields taken, and that was how wealth moved across the earth. The division of labor rewrites the equation. Mises generalized the point into what he called the law of association: even the man better at everything gains by trading with the man worse at everything, so there is no person on earth it does not pay to cooperate with.

Under that law the foreigner producing and trading is worth more than the foreigner robbed once and ruined, and this is why the deepest classical liberals were also the hardest enemies of war and empire—not out of softness but out of arithmetic. The market is the great solvent of the reasons men have butchered one another: it makes the stranger more useful as a partner than as a corpse. It does not turn us into angels. It makes cooperation pay better than conquest and then lets ordinary self-interest accomplish what sermons never could—giving a man a living stake in the safety and freedom of people he will never meet.

Rand carried the argument to its constitutional form. A civilized society, she wrote, is one that has taken physical force out of human relationships altogether, and the American Revolution mattered because it was the first attempt to build a nation on that bar—to put the agreement above the club. Capitalism is the ban on force gone to work: millions of people, every hour, choosing persuasion over the fist, and being fed for it. Freedom and capitalism are inseparable because they are one choice made in two places—the courthouse and the shop.

❧

The arguments against all of this are old and many. Every one of them, followed to its root, is the same proposal: switch off the correcting signal and put a command in its place—solve by force a problem the market was already solving by discovery. That is not an incidental flaw in the anti-market case. It is the whole case.

Take the strongest of them, the objection about the costs a price does not capture. When a factory fouls a river, the harm is real and the price of its product records none of it, and here, surely, is a true failure of the market. But look again at what has gone wrong. The river was fouled because it belonged to no one—because there was no owner to charge for its ruin, no price on the damage, and therefore no signal and no correction. The failure is not that a market operated. It is that no market was permitted to form, because the thing being destroyed had been left outside the web of ownership and exchange that prices everything inside it. The cure is to run the correcting machinery out until it reaches the river—to give the river owners and prices and defenders—not to smash the machinery in the name of the one thing it was never allowed to touch. What gets called a failure of the market is, repeatedly, the name for a place where the market was forbidden to go, its absence then held up as proof that it should never be let in.

When a better product destroys an older industry, real people in a real town lose the work their lives were built around. This is true. It is also *only the seen half of the story*. The unseen half—the half Sowell spent a career, after Bastiat, teaching economists to see—is the far larger number of people, most of them strangers and many of them not yet born, who are lifted by the very process that closed the mill: the cheaper goods that leave a poor family more at the end of the month, the labor and capital freed to build what does not yet exist, the whole onward motion of correction that the old industry's survival would have frozen in place.

To halt that motion in the name of its visible casualties is to sacrifice the vast invisible many to the few one can photograph—and to do it, necessarily, by force, since the only way to stop willing people from buying the better thing is to stand between them with the law. The compassion is not in doubt; the arithmetic is upside down. Ask the question Sowell taught a generation to ask of every warm-hearted fix—*and then what happens?*—and the mercy curdles, a stage or two downstream, into a poverty deeper than the one it hurried to relieve.

❧

On the twenty-fifth of March 1911, fire caught in the Triangle Shirtwaist factory in New York, and inside eighteen minutes a hundred and forty-six workers were dead—most of them immigrant girls, some of them fourteen. They died because the owners had kept the exit doors locked to stop the girls slipping out with unbought thread, and when the flames came the locked doors held. Some jumped from the ninth floor rather than burn, and the crowd in the street below watched them break on the pavement. This is the picture the whole case against the market is built on, and it has every right to be. No honest word about the freedom to trade can be spoken over those bodies without stopping at them first.

Look hard, then, at what killed them, because the lesson usually drawn gets the cause wrong. It was not the wage the girls had agreed to; a low wage, freely taken, does not lock a door. It was the locked door—a lethal hazard fastened onto people who had no way, when they took the work, of knowing the exits would be locked behind them. That is not a bargain. It is the oldest wrong there is, one person forcing a deadly risk onto another without their knowing, and every account of a free society worth the name has always called it a crime and set the law on it. The correcting order does not ask that such a thing be borne; it asks that it be punished, ruinously and without exception, until no owner on earth would run the sum and lock the door. What failed at Triangle was enforcement: a wrong the law already forbade went unpunished until the girls were dead.

There is one more turn, and it sets the comparison back on its feet. The mind puts the 1911 sweatshop beside the clean and daylit workplace of today and recoils, rightly. But that was never the choice anyone faced in 1911. The true comparison is between the sweatshop and the thing the worker had walked away from to reach it—the worn-out farm in Calabria or Galicia, or the peasant hovel outside Minsk that had been starving her family for a hundred years.

Set against the workplace of the present, the sweatshop *is* a horror. Set against the field of the past, it was, for the people who ran to it, the better bargain—and the rising wages that would soon empty such shops of their children were already climbing while the reformers wrote. The men and women who carried cameras into those mills and would not look away saw true suffering and reported it truly, and we owe them for it. Where they went wrong was the diagnosis: they traced the harshness to the freedom that was, year by year, undoing it, and reached for a cure that would have frozen the climb halfway. The suffering was real. It was the last of the old poverty draining out, and the factory was the drain, not the wound.

When rents climb, the rising price is doing its harsh and honest work: shouting at builders to build and warning tenants that shelter has grown scarce. The humane-looking answer, tried in a hundred cities, is to forbid the rise by law—to cap the rent, which is to say, to jam the signal by force. And every time, the same sequence follows, because a signal that cannot move cannot correct. New building stops, because no one will sink a fortune into what he is forbidden to price. The apartments that already exist are no longer steered toward those who need them most but frozen where they sit, held by whoever got there first. The shortage the cap was meant to end hardens into a permanent famine of housing, and the city fills with people hunting an apartment the law has made it pointless to build. The control did not fail to fix the problem. It manufactured a worse one, by the short, straight, wholly predictable road of turning off a correction and handing down an order in its place.

Rent caps, wage floors that price the least-skilled clean out of a job, whole economies planned into starvation—they are one thing in many coats, force laid across a signal, and each brings back, with the fidelity of a natural law, the exact ruin it swore to prevent. Mises traced the spiral in advance: *each control creates the distortion that becomes the argument for the next control, the failure of every dose read as proof that the dose was too small.*

The serious objection gathers these into a single hope: if the price is only a computation, and computation is exactly what our machines now do at superhuman scale, then perhaps the thing Mises said was impossible has been waiting for enough processing power. Perhaps a big enough computer, fed enough data, could finally calculate the plan—could be the central mind the twentieth century lacked, and run the economy better than the blind scramble of the market ever could. It is the most sophisticated form of the dream of planning, and it fails for a reason that is subtle and complete.

The knowledge a market runs on does not exist in a form that could be fed to any computer, because it does not exist until the market elicits it. You do not know what you would pay for a thing until you are standing in front of it with your own money on the line; the entrepreneur does not know his guess was wrong until he has staked something real on it and watched reality answer. The data the planner would need are not lying around uncollected. They are generated, freshly, moment by moment, by the process of free people choosing—and there is nothing to feed the machine, because the numbers do not exist in advance of the choosing that creates them. You cannot compute your way around a market for the same reason you cannot compute in advance the result of an experiment you have not run. The market is not a calculation waiting to be done faster. It is a discovery procedure, and what it discovers does not exist until it discovers it.

The pencil on the desk and the fed city are not small things, and they are not, in the end, about wealth. They are the correcting animal doing at the scale of a species what it first did in a crib: throwing guesses at a world too large for any one of its minds to hold, and keeping what survives the test of reality. The market is the widest circle of cooperation human beings have ever drawn—wider than the family, wider than the nation. It binds a Malaysian rubber tapper to an Oregon logger to a child buying a pencil, none of whom will ever meet, none of whom need agree on anything, each of whom is made better off by strangers acting, without knowing it, on his behalf. It is the reach past the tribe made into a permanent machine. And when it is smashed, in the name of a plan that will do better, what returns is not a kinder order but the oldest darkness there is: a system that cannot see, ruled by people who have mistaken the abolition of the correcting signal for the conquest of the problem it was correcting.

The case against the market is always the same case: curtail the correcting of error and do it by force. And those two things—the sealed error and the gun—are the marrow of evil. The quarrel with the freest cooperation our kind has ever built is not one argument among many about economics. It is the definition of evil, speaking the language of compassion.

The order that comes out of all this un-commanded striving—the fed city, the stocked shelf, the pencil that no one designed—has a quality that is hard to name and impossible to miss once you have seen it. It is a kind of beauty: an intricate, functioning, self-repairing order that no mind planned and no mind could have planned, arising from the free play of millions of small corrections. And the fact that human beings find such order beautiful—that we are moved by it, that a working thing strikes us as somehow right in a way that goes beyond its use—is itself a clue about what we are.

CHAPTER TEN

Why We Make Beautiful Things

Forty thousand years ago, in a cave in what is now southern Germany, someone sat down with a mammoth's tusk and a stone blade and carved an animal that never existed.

It has the body of a man and the head of a cave lion, and it stands about a foot tall. To make it, a person in an ice-age world of relentless scarcity—where every hour not spent hunting or gathering or keeping warm was an hour gambled against starvation—turned away from all of it and gave hundreds of those hours to a figure of pure imagination. You cannot eat it, wear it, or hunt with it. It kept no one alive. It is the oldest known sculpture on Earth, and it is a lie: a careful, patient, beautiful lie about a creature that never walked the world.

The carver was not copying the world, the way the earliest painters copied bison and horses. He was taking the world apart and building a new one—a lion here, a man there, welded into a thing that had never been—which is the same power that lets a mind picture the winter before it comes and the tool before it exists. The oldest art we possess records nothing anyone had seen. It is proof of what a mind had learned to imagine.

An archaeologist once sat down with the same materials the carver had—a length of ivory, blades of knapped flint—and set out to reproduce the figure by the original method. It took him something on the order of four hundred hours. Four hundred hours, in a world where the difference between a full belly and an empty one was measured in daylight, spent by someone on a thing that fed nobody and killed nothing, for no reason anyone has ever been able to reduce to survival. Whatever drove that person was strong enough to outweigh the arithmetic of staying alive.

The figure came out of the ground in pieces—hundreds of slivers of ivory dug from the cave floor in 1939, days before a war swallowed everyone's attention, then boxed and shelved and half-forgotten for decades before anyone fit them back together into the lion-headed man. Even then the reconstruction was partial; a fresh excavation in 2013 sifted still more fragments out of the old dirt, and the figure grew more whole, forty thousand years after its maker set down the blade. We are, in a small way, still finishing his work.

And it was no fluke of one strange genius. The same cluster of caves gave up the oldest musical instruments ever found, flutes cut from the hollow wing-bones of birds and tuned forty thousand years ago, so that alongside the first sculpture sits the first music. Half a world away, on the Indonesian island of Sulawesi, a wild pig was painted onto a wall in red ochre at least forty-five thousand years ago, older than anything in Europe, and beside it strides part-human, part-animal hunters who also never existed. Older still, in a South African cave called Blombos, someone scratched a careful cross-hatch pattern into a chip of ochre seventy thousand years ago—no animal, no scene, just pattern for its own sake, made tens of thousands of years before the first cave paintings by a mind that already wanted to press an order onto a surface that had none. Wherever the bones of modern humans surface, the art surfaces with them, arriving the moment we do.

The caves themselves keep raising the bar. When explorers lowered themselves into a sealed chamber in southern France in 1994, they found a gallery of horses and lions and woolly rhinoceroses drawn thirty-six thousand years ago with a command that stopped the scholars cold—animals shaded to bulge from the curve of the rock, caught mid-stride, a whole pride of lions rendered in a few unhesitating lines. These were the work of people who already commanded the whole craft, at the dawn of the species. Nothing in them is tentative. Down at Lascaux, found by four boys and their dog chasing the lost pet into a hole in the ground in 1940, the great bulls run across the ceiling in ochre and charcoal with a life that the modern painters who came to see them recognized at once as the real thing, fully formed, seventeen thousand years before there was a Louvre to hang it in.

When the proof first came out of the ground, the experts refused it. In 1879 a Spanish nobleman exploring a cave called Altamira was called over by his small daughter, who had wandered into a low chamber and looked up—*look, Papa, oxen*—at a ceiling crowded with bison painted in red and black, alive with a skill no one expected. He announced them as the work of Ice Age hunters, and the scholarly world laughed him down and accused him of hiring a forger, because the men who studied prehistory could not make themselves believe that primitive savages, as they thought of them, had hands and eyes and souls equal to that ceiling. It took twenty years and the discovery of painted caves across France to force the retraction. The nobleman died before it came, dismissed to the end as a fraud for having told the truth. The disbelief is the tell. Confronted with beauty made forty thousand years ago, the modern mind's first instinct was to deny that the ancient mind could have made it—to protect the flattering story that the capacity for beauty was something we acquired late, along with cutlery. We did not acquire it late. We have never been without it.

Music is the strangest of all these gifts, because it is a picture of nothing. A cave lion is a picture of something; a melody points at no object in the world, and yet a run of notes in the right order can raise the hair on your arms, flood your eyes, move a room of strangers to sway in time. Darwin, who could account for so much, confessed himself beaten by it—he ranked the human love of music among the most mysterious faculties we own, useless on its face for survival and universal in a species that agrees on almost nothing else. Every culture ever found has made it. None has ever explained why an arrangement of sound with no meaning can break a heart.

Steven Pinker gave it its most memorable name: art and music are *auditory cheesecake.* Cheesecake delights us because we evolved a hunger for fat and sugar in a world where both were scarce and lifesaving; the cheesecake itself does nothing for survival, it merely floods circuits built for other work. Art, on this reading, plays the same trick on the eye and the ear—a pleasure technology that reaches in and presses buttons shaped by natural selection for the grim business of staying alive, a confection our senses cannot help enjoying and our species could lose tomorrow with no cost to its fitness. The mind, on this account, is a bundle of problem-solving machinery built to hunt and mate and outwit rivals, and art is what happens when that machinery, at its leisure, discovers it can tickle itself—a set of super-stimuli lighting up the pleasure circuits without doing any of the work those circuits were installed to do. Charming, harmless, and beside the point.

A luxury.

A second answer says beauty is a display. Geoffrey Miller argues that art is the human peacock's tail: expensive, difficult, impossible to fake, and therefore an honest advertisement of a quick brain and time to burn. The cave painter, like the peacock laboring under its plumage, is broadcasting to prospective mates—*look what I can afford to waste.* On this account the drive to make beautiful things is the drive to be chosen, dressed in pigment, and the finest art is the most extravagant possible proof that its maker carries genes worth having. The logic is the peacock's exactly: a signal can be trusted only when it is costly, since only the genuinely gifted can afford to pour so much into something so useless—so the wastefulness of art, the years and the obsession and the refusal to stop at good enough, becomes the guarantee that the display is honest. Miller turned a clean edge against Pinker, one that cuts. Cheesecake, he pointed out, explains why we enjoy art, which is the pleasure of the audience, and says nothing about why a person would bleed to make it, which is the compulsion of the maker.

He had explained the groupies and left out the rock star.

There is a third answer, closer to the truth than either, and it starts from a strange fact about what human beings find beautiful. Ask people across wildly different cultures to describe or choose an ideal landscape and they converge, eerily, on the same scene: open grassland dotted with trees, water in the middle distance, a path or a river leading the eye toward a horizon, some animals, a sheltering rise to look out from. It is, near enough, the African savanna our species grew up on—the terrain a vulnerable ape would read at a glance for water, escape, and game. Something in our sense of beauty is tuned to the shape of a world that once kept us alive.

Two artists once ran this as a stunt that turned serious, polling people across a dozen countries on what they wanted to see in a painting and then dutifully painting the averaged answers. From Kenya to China to Iceland the reply came back almost identical: a blue landscape, water, a scattering of trees, a few animals, human figures in the middle distance, an open view toward a horizon—the calendar scene, the savanna in disguise, wanted by nearly everyone nearly everywhere. Taste, which we treat as the most private thing we own, turned out to have a floor beneath it that was laid down before we were human.

Ellen Dissanayake built the fuller version of this: that art is the human habit of *making special,* of taking the ordinary object and the ordinary act and elaborating them past all necessity—the pot that could have been plain and is glazed, the story that could have been a report and is sung—and that this making-special is woven through everything we most need to hold a group together, the wedding and the funeral and the harvest. Beauty, on this reading, is one of the tools of survival, the thing that binds a people to each other and to the world they must live in.

Miller's peacock is a male bird, and its tail grows for one season and one end. Women make art. Children make art. People make it alone, in private, for no eyes but their own, and go on making it long after every mate is won or lost or never sought. The old man whittling on a porch with nobody watching, the widow who keeps her easel by the window, the child who fills notebook after notebook with worlds no adult will ever be shown—not one of them is courting anyone, and not one of them is merely idle, and all of them do it anyway, because the doing answers something the theories have not managed to find. And Pinker's cheesecake is a thing you reach for when you are safe and fed and idle—which happens to be the one condition under which the greatest art in human history was not made.

None of the three theories is exactly wrong. Art does signal fitness; it does please the senses; it does bind a people together. Each caught a real piece of the thing and mistook its piece for the whole. What all of them miss is the compulsion that outlasts every use—the making that goes on when there is no mate to impress, no leisure to fill, no group left to hold together, when there is nothing left at all. To see that, you have to go to the one place where every instrumental reason for art had been stripped down to nothing, and the art was made anyway.

❧

In the ghetto at Terezín, in occupied Czechoslovakia, the Nazis held some fifteen thousand children on their way to Auschwitz. Fewer than a hundred and fifty survived. They were hungry in a way that resists the page—survivors remembered sucking their coat buttons to fool their stomachs—and most of them were already condemned, though how much each child understood of that no one can say.

Among the prisoners was a painter named Friedl Dicker-Brandeis, trained at the Bauhaus under some of the great teachers of the age, and she did an extraordinary thing. She ran drawing classes. She had come to believe, over years of teaching, that making art did something to a person no circumstance could confiscate—that a child bent over a drawing was, for that stretch of time, free in the one place the guards could not reach, and that this inner freedom was worth teaching even to children the world had already sentenced.

In secret, on smuggled scraps of paper, she taught the children of Terezín to make pictures, and she gave them a prompt. *Paint the place where you wish you were now.* They painted butterflies. They painted flowers, and windows with light coming through, and the roads that led home. Before she was loaded onto the transport that killed her, Dicker-Brandeis packed two suitcases with roughly four and a half thousands of these drawings and hid them. The drawings were found after the war. The children, mostly, were not. The collection that survives is the largest body of children's art the Holocaust left behind, and one of the poems saved with it gave it its name: the children, it said, never saw another butterfly.

Set that beside the two theories and watch them come apart. There was no mate to court in Terezín, and no fitness to advertise to anyone who could reward it. There were no spare calories; every one was counted, and there were too few. By the cold arithmetic of the cheesecake and the peacock, not one of those children should have lifted a pencil—every scrap of strength should have gone to the single project of living one more week. Instead, starving and doomed and running out of days, they made beauty. Small, bright, useless, unbearable beauty, because the making of it is one of the things a human being is, and the making does not switch off even when everything else has been stripped away. The drawings did not save the children. That is not what they were for, and to pretend otherwise would be an insult to the dead. They are evidence—of exactly what was being murdered, which was not only fifteen thousand lives but the particular thing that a human being is, the thing that reaches for beauty even at the bottom of the world.

And it was not only Terezín. The same impulse surfaces wherever people are driven to the edge and somehow still have their hands: the figures prisoners scratch into cell walls, the songs that arose out of the slave quarters and the labor camps, the poems written in the trenches in the lull between shellings, the drawings made in hospices by people with weeks to live. No one ever had to be taught to do this. When everything instrumental has been stripped away—freedom, safety, future, sometimes the near-certainty of tomorrow—the making of something beautiful is among the last things to go, and often the very last. A luxury does not behave like that. A necessity does.

Ayn Rand defined art as a re-creation of reality that is selective, shaped by the artist's deepest judgments about what matters, and selective is the working word. The carver in the cave chose the lion, chose the man, and left out everything else on earth, and the choosing is the statement. A creature that steers its life by abstractions—justice, home, the future, the good—needs somewhere to see them, because an abstraction cannot be looked at. That, on Rand's account, is what art is for. It takes the widest things a mind lives by and hands them back as a single object the senses can seize, a whole appraisal of existence compressed into a thing; she called that underlying appraisal *a sense of life*, the wordless sum a person carries of what the world is and what can be hoped from it. The hush in front of the painting is one sense of life recognizing another—or colliding with it.

And she said one more thing, which Terezín had already proven. Art, Rand wrote, gives a human being *emotional fuel*—the experience of feeling, for an hour, what it would be like to live in the world one's values point toward. Read the teacher's instruction again with that in hand. Paint the place where you wish you were now. Friedl Dicker-Brandeis was not distracting the children of Terezín. She was fueling them—handing each child an hour inside the world her values named, the window with the light coming through, the road that led home—and the drawings were not a refuge from the fight to stay human. They were the fight, provisioned. The old man whittling on the porch with no one watching is running the same operation in peacetime: contemplating, in a shape his own hands made, the way he holds the world. No mate, no audience, no calorie is involved, and no mystery remains once the making is understood as a mind keeping its own values in sight.

The question was wrong from the start. Not why do we make beautiful things, as though beauty were a decoration we add once the serious work is finished. The making of beauty is the serious work, and to see why, you must leave the cave and the ghetto and walk into a physics department.

❧

In 1928 a young Englishman named Paul Dirac was hunting an equation to describe the electron. He was not trying to match an experiment. He was trying, by his own account, to make the mathematics beautiful—to find the form with the right symmetry, the cleanest fit, the shape that felt true—and when he found it, his equation did something no one had asked of it. It insisted that for every particle there must exist a mirror twin of opposite charge: an anti-electron, a thing nobody had ever seen or imagined. Dirac worked in a way that unnerved his colleagues—sitting and turning the symbols over, he said, chasing not a prediction but a certain rightness in their form, on the settled conviction that the equations describing the deep structure of the world would, once they were correct, be beautiful, and that an ugly line of mathematics was reason enough on its own to suspect it was false.

Most physicists thought the extra solutions were an embarrassment to be explained away. Dirac trusted the equation over the common sense. Four years later a young American, sifting cosmic rays in a cloud chamber, found the anti-electron sitting there exactly as the beautiful mathematics had demanded, and Dirac had his particle and, soon after, his Nobel Prize. He defended the method to the end of his life. It is more important, he said, to have beauty in one's equations than to have them fit experiment.

He was not alone in it. Feynman said you could recognize truth by its beauty and simplicity. Einstein chose the elegance of general relativity over the settled physics of centuries and turned out to be right. Keats had put the whole creed in three words a hundred years before any of them: beauty is truth. The people who go deepest into the structure of reality keep reporting the same thing: that the true answer, when they finally reach it, is beautiful—that the beauty is the feel of the answer being right.

There is a deeper riddle folded inside this, one the physicists themselves puzzle over: that the universe should be describable by beautiful mathematics at all. Nothing obvious required the deepest laws of reality to be elegant. They could have been an ugly sprawl of arbitrary rules with no pattern a mind could love. And yet, over and over, the equation that turns out to be true is the one a trained eye calls simple, symmetrical, clean. Eugene Wigner named this the unreasonable effectiveness of mathematics, and no one has fully explained it. The world appears to be built along lines a mind can find beautiful—which is either the largest coincidence there is, or a clue about what minds and worlds have in common.

Here is what the cave carver and the physicist share, and it is the hinge of the whole matter. *Beauty is the sensation of a mind finding the true shape of something.* The right line of a running horse on a cave wall; the arch that stands because its curve is correct; the equation whose symmetry shuts like a lock; the butterfly a dying child makes cohere on a scrap of smuggled paper—each is a single event felt from the inside, a mind pressing order onto chaos and knowing, by a rush of pleasure, that the order is real. We are built to feel that rush, because a creature that survives by finding the true shapes of things needs a signal to tell it when it has succeeded, and the signal has to be strong enough to pay for four hundred hours of carving and to reach a starving child in a death camp. Evolution gave the search for truth a reward, and the reward is the experience of beauty.

This is why the pull runs so strong and so strange, why it can feel almost like revelation, why people reach for the language of the sacred to describe a great proof or a great painting. They are not overstating it. They are reporting, accurately, the sensation of a mind clicking into alignment with the way something is—the same click, scaled up past all proportion, that a child feels when the last piece of the puzzle drops home. Beauty is what correctness feels like from the inside. And because the search for the true shape of things is the one activity this animal was built around, the reward for it had to be among the strongest feelings we have.

The shiver a piece of music sends down the arms, the sting behind the eyes at the line of poetry that lands, the involuntary hush in front of a painting that is somehow exactly right—each is the nervous system registering a hit, the same recognition the maker felt, arriving secondhand. We are wired to feel it in the making and in the meeting both, because a signal worth having had to be one that neither the discoverer nor the witness could talk themselves out of.

Which means the beauty was the same fire all along. The reach that remakes the world is the reach that turns chaos into form and finds the form beautiful: one faculty, one creature, aimed now at survival and now at meaning, and unable, at the deepest level, to tell the two apart. To make a thing true and to make a thing beautiful are the same act, performed on different material.

Art and science are taught as opposites, one belonging to the heart and the other to the head—which is why it startles to hear a mathematician and a poet both call an answer lovely, without irony, and mean the identical thing.

It also crosses every border a culture ever drew. A person raised on nothing but Western music can be stopped in their tracks by a raga or a call to prayer; an eye trained only on European painting still falls quiet before a Persian miniature or a wave by Hokusai; a proof that strikes a mathematician in Kyoto as elegant strikes one in Cairo the same way. We argue without end about tastes, and beneath the arguing runs a shared floor—a species that agrees, across every ocean and every century, that some made things have found the true shape and others have missed it. That agreement is the faculty showing through the costume of the culture.

The signal is not flawless. Beauty can lie. Dirac himself, chasing symmetry, predicted a particle called the magnetic monopole that a century of searching has never found. The old astronomers thought the heavens moved in perfect circles because the circle was the most beautiful figure, and the sky patiently refused them for two thousand years. Sabine Hossenfelder, in *Lost in Math*, argues that the modern hunt for beautiful theories has led her field into a wilderness of elegant mathematics with nothing underneath it. The danger is real: a mind that has learned beauty tracks truth can fall in love with an elegant idea and defend it long after the evidence has quietly left the room. Beauty is a compass. It tells you which way feels true, and mostly it is right, and every so often it points with perfect confidence off the edge of a cliff. But that beauty sometimes misfires is the feeblest possible complaint against it, because a mind that felt no such pull at all would have no way to tell a true shape from a false one. It would build nothing and discover nothing. The caves would be bare.

There is a question survival never answers: what any of it is for. No one on the last day of a life asks for one more afternoon of the merely useful—one more meeting, one more errand, one more unit of bare survival. They ask for the people they love and the things that moved them: a piece of music, a stretch of coast at evening, a poem learned in childhood, a face. Watch, too, where people spend the freedom the useful buys them. They cross oceans to stand in front of one painting or under one particular sky; they fall silent, without being told to, inside a great cathedral or at the rim of a canyon; they mark every wedding and every funeral—every hinge of a life—with music, because the plain words will not carry the weight alone. A species that files beauty under luxury has arranged its leisure and its wealth and its deepest occasions around getting more of it. The useful is the thing we do so that the beautiful can happen. Survival is the setup. Beauty, and the love that is its twin, is the payoff.

The child in the ghetto, the hunter carving a lion in the dark, and the physicist reaching for antimatter are one kind of creature doing one thing, in the only way it knows: reaching past the world as given toward the truer shape of a better one and knowing it has arrived by the beauty of what it finds. A human being who made no beautiful thing would be the one who needed explaining. The rest of us are only being what we are.

It is one power under all of it—the coat pulled from the fox, the sea walled out, the plague run to ground, the stranger made into a spouse, the cathedral, the lion carved in the dark, the equation that reached out and touched antimatter before anyone had seen it. The same fire that lets a mind leave the world as it is and remake it nearer the true shape it can see is the most powerful thing on Earth, and the most dangerous. And it was, the old story says, stolen.

CHAPTER ELEVEN

Prometheus Unrepentant

There is a mountain and a figure chained to it. Each morning an eagle comes and eats his liver; each night the liver grows back, so the feast can begin again at dawn. In the story this goes on for thirty thousand years. The prisoner is a god named Prometheus, and his crime was a gift: he stole fire from heaven and handed it to human beings, and for that the ruler of the gods pinned him to the rock and sent the bird.

Prometheus—the name means *forethought*—was, in the older versions, both the maker of mankind and its champion, and the fire he stole was never only the flame. In Aeschylus it stands for everything he smuggled down to us with it: writing, number, medicine, the reading of the stars, the working of metal, the taming of the horse. He gave us, in one word, *technology*—the entire human power to reach into the world and remake it nearer to what we want.

Before that flame we were a middling, frightened animal on a plain full of things with better teeth: slow, soft-skinned, easy meat after dark. The fire changed our standing in the world almost overnight. It pushed back the night that the great predators owned, hardened the point of the spear, warmed the ones who would otherwise have frozen, and turned a terrifying world into something a naked ape could begin to hold at arm's length. Before we had it, we were prey. After we had it, we were the thing the rest of the world learned to fear. He tricked the gods twice to do it, once over the division of a sacrifice and once over the fire itself, which he carried down the mountain hidden in the pith of a fennel stalk. And the gods answered the double theft with a double punishment, one for the giver and one for the given.

For Prometheus, the rock and the eagle. For mankind, Pandora: the first woman in the story, sent down carrying a sealed jar, which she opened, loosing into human life every evil it had lacked before—labor, sickness, war, old age, death. One thing alone stayed inside when the lid came back down. *Hope.*

The fire arrived with a bill attached. What the myth cannot decide is whether the bill means the gift was a crime.

Its two oldest tellers disagree, flatly. Hesiod, who set it down first, takes the side of the gods: Prometheus is a thief, human misery is the wages of his theft, and Zeus is only justice doing its work. Aeschylus, two centuries on, turns the whole thing over. His Prometheus is the hero of the play and Zeus the tyrant; the Titan hangs on his crag unbroken, refusing to repent, throwing his defiance in the face of the power that tortures him. The same story, the same fire, and two opposite verdicts on whether we should kneel in gratitude or in shame.

We have never stopped arguing it. When the Romantics picked the myth back up, they split along the very seam that separates Hesiod from Aeschylus. Percy Shelley wrote *Prometheus Unbound*, in which the fire-bringer is loosed from his rock and mankind loosed with him—liberation, ascent, glory. Mary Shelley wrote the other half. Her novel about a man who steals the divine trick of making life and is destroyed by the thing he makes, she subtitled *The Modern Prometheus*. We call it *Frankenstein*. The two books are the two verdicts, and between them they hold nearly everything we still say about our own cleverness: that it will save us, that it will damn us.

The West did not even need the Greeks to think this thought, because it had already thought it once, in a garden. There is a tree in that garden whose fruit is the knowledge of good and evil, and a single rule against eating it, and a serpent who promises that on the day you eat it your eyes will open and you will be as gods. The two humans eat. Their eyes open. And for the crime of reaching past the animal's blessed ignorance into knowledge, they are cursed with exactly Pandora's inventory—pain in childbirth, toil in a hostile ground, the sweat of the brow, and death at the end of it—and driven out of the ease they were born into, east of Eden, with an angel and a flaming sword set behind them so they can never go back. Two founding stories stand at the head of our civilization, one Greek and one Hebrew, and they tell the same tale with the same moral: the knowledge that lifts us toward the gods is a theft, and the theft is paid for in suffering, and we would have been better left in the dark.

And it is not only ours. Nearly every people that ever gathered around a fire has told a story about where the fire came from. The Polynesians say the trickster Māui dragged it out of the fingernails of an underworld goddess. The Norse gave the stolen cunning to Loki, who is bound, like Prometheus, under torment for what he set loose on the world. Peoples who never traded a single word with one another arrived, on their own, at the same uneasy tale: that the power which lifted us above the animals was taken and not given, that it came wrapped in danger, and that some god somewhere had not meant for us to have it. The unease about our own reach is not a quirk of one tradition. It is very nearly a human universal—a measure of how deep the fire runs in us, and how old the temptation is to be sorry we hold it.

Both stories are wrong. And the myth turns out to be more literal than any of its tellers could have guessed.

❧

Roughly two million years ago, one of our ancestors began to cook. We cannot fix the date from the hearths, because ash does not survive the eons, but the bodies keep the record the fires could not. Around the time fire enters the human line, the animal is remade: the gut shrinks, the jaw and the teeth grow smaller, and the brain begins its long swell toward the organ behind your eyes. The anthropologist Richard Wrangham supplied the mechanism. Raw food is expensive—the great apes spend as much as seven hours a day chewing it and burn a large share of what they swallow just breaking it down. Cooking does the breaking down in advance, outside the body. It pours into the eater the calories that the raw root and the raw carcass had kept locked away, and it hands back the hours the chewing used to eat. And the brain is the most expensive organ we own; it runs on something like a fifth of everything we burn, all day, asleep or awake. No ancestor could have paid for a brain like ours out of a raw diet. Fire paid for it.

The date is disputed. Hard evidence of tended hearths runs back only some hundreds of thousands of years, well after the brain had begun to swell, and serious researchers still argue over the order of cause and effect and whether meat or cooking did the most. But no one disputes the destination. Whenever it happened, the one animal that learned to master fire is the one animal that grew a mind, and the two facts have never been found apart.

The organ that would one day design the reactor and the requiem was itself built by fire—grown, cell by cell, on the surplus that only cooking could release. The taming of the flame reached inside and enlarged the thing we think with. The fire that split the atom first cooked the meat that grew the brain that split it. Prometheus's gift is not a symbol standing in for human power. It is the physical, dietary cause of the human brain, and there is no version of our species that declined it and stayed our species. To wish we had never taken the fire is to wish away the very thing that is doing the wishing.

And the flame did more than enlarge the brain; it drew the circle. A hearth has to be tended in turns and shared, and around it the whole shape of the day reorganized itself—food carried back to a common place instead of consumed where it was found, the labor of the band divided, the evening opening up for the first time as a stretch of light that was spent neither hunting nor chewing but sitting, together, in the glow. The cook-fire is the oldest social institution we have. Long before it was the furnace and the engine and the reactor, it was the thing we gathered around, and the gathering is as much its gift as the calories. To take the fire was to become, in one motion, both cleverer and less alone.

When people speak of the curse of Prometheus—the price of our cleverness, the shadow over our power, the worm in the fruit of the tree—they are nearly always fusing three separate things into a single dread, and the three do not belong together. There *is* a price. There *is* a weight. And there *is* a betrayal.

Take them one at a time.

The price is the plainest of the three. *Fire is work.* It must be gathered and tended and guarded against going out, and everything downstream of it—the planted field, the forge, the engine, the operating theater—is labor and vigilance and unending care. Pandora's jar, read without melodrama, is largely just this: the same fire that lets us reshape the world binds us to the ceaseless upkeep of a made world, with all the toil and worry that keeping it demands. The curse laid on Adam is the identical curse, spelled out in agriculture—the ground will fight you, and you will eat only what you have sweated for. That is real. It is also the cost of the gift, the way any powerful tool is a duty as much as a possession. No one who has grasped the bargain would hand the fire back to be free of the chores. The price is paid, and paid gladly, because what it buys is everything else.

The weight is the serious one.

At half past five in the morning on the sixteenth of July 1945, in a stretch of New Mexico desert, the first nuclear weapon exploded. The man who had led the work of building it, Robert Oppenheimer, watched the fireball climb and, by his brother Frank's account, said only two words at the time: *it worked.* The vast sentence everyone now remembers—*now I am become Death, the destroyer of worlds*—he supplied two decades later, in a television interview, reaching back into the Bhagavad Gita to name what the desert had felt like. Three weeks after the test, on the sixth and the ninth of August, two of these weapons were dropped on two cities in Japan, and something on the order of a hundred and fifty thousand people—overwhelmingly civilians, a great many of them inside a single morning—were killed. Some were gone in the first flash of light. Others died for years afterward, of the fire's slower poisons.

Five months earlier, over a single night in March, ordinary firebombs had burned as much of Tokyo and killed as many people, something near a hundred thousand, with nothing more exotic than jellied gasoline. The century was already fluent in mass death; the physics only concentrated it. What the atom changed was the ratio: one bomb, one plane, one morning, where the same slaughter had taken a thousand planes before. And the men who made it were not monsters. Oppenheimer said the most damning and most honest thing about the whole effort when he called the problem *technically sweet*—that when you see something technically sweet you go ahead and do it, and you argue about what it means only after you have had your technical success. That is the fire in its purest and most dangerous form: the pull of a beautiful problem, so strong that the hands have solved it before the conscience has finished asking whether they should.

But fascination did not drop the bombs, and neither did cruelty. In the summer of 1945, the men deciding were staring at a short menu of horrors. The invasion of Japan was planned and dated—Kyushu in November, the Tokyo plain in the spring—and the casualty figures put before Truman that summer ran from a quarter of a million Americans at the low end to a million at the high, with Japanese losses reckoned at least their equal. Okinawa, just concluded, had cost on the order of a quarter of a million lives in twelve weeks, a great many of them civilians, and the code-breakers were watching the invasion beaches of Kyushu fill with divisions. The services had already minted the medals: roughly half a million Purple Hearts, struck in advance for the invasion—a stockpile so large that every American wounded in every war since, through Korea and Vietnam and into this century, has been handed a medal made in 1945 for a battle that never happened.

Nor did waiting cost nothing. In the lands Japan occupied, the war was killing on the order of a hundred thousand civilians every month it continued, and the blockade was walking Japan's own population toward mass famine. The choice was never the bomb or peace. It was the bomb, or the beaches, or the hunger, or a settlement that left the regime armed with its conquests in hand—and the men who chose believed, on the evidence in front of them, that they were choosing the smallest mountain of dead.

That arithmetic sits inside a moral fact that must not be blurred. *There were those who started that war and those who fought to end it*, and the weight of the dead, the cities included, belongs on the ledger of the starters. A regime that launches a war of conquest and then refuses every exit from it has built all the doors, and the men who finally chose the least terrible one were not its partners in guilt. They were its victims' last resort. Historians still argue the margins—Tsuyoshi Hasegawa has made the case that the Soviet declaration of war, landing between the two bombs, did more than the fission to break the cabinet in Tokyo. What it does not touch is the structure. The deciders were performing, under the worst conditions it has ever faced, the correcting animal's own act: weighing guesses about an unknowable future, choosing, and letting reality grade the choice. The gravest use of the fire ever made was not the fire slipping its leash. It was men holding the fire, trying to end the burning.

This is the fire too. The same power that cooked the first supper and grew the first human brain and raised the cathedral at Cologne also made the thing that unmade Hiroshima.

Oppenheimer looked at what they did for the rest of his life. *The physicists have known sin,* he said afterward. The book that tells his life is titled, with great precision, *American Prometheus,* because the old pattern ran its course to the end: the man who carried the fire down was punished for the carrying. In 1954 his own government stripped him of his security clearance in a hearing designed to break and humiliate him, and the fire-bringer was chained to his rock while the country he had armed sent the eagle.

And yet the same man who said the physicists had known sin also said a thing the guilt hurries past. He said he carried no weight on his conscience about the science itself—that what had been found was a fact about the world, and that the use made of the fact was the business of governments, not of the men who had found it. He was drawing a line, and the entire question turns on that line.

The fire that leveled a city is the very same fire that ends the plague, lights the ward, and warms the child through the winter. There are not two fires, a good one and an evil one, such that we might keep the first and hand back the second. There is one fire. The knowledge that powers cities is the knowledge that builds the bomb; the chemistry of the cure is the chemistry of the poison; the reach that can remake the world can unmake it, of necessity, because it is a single reach. No one has ever found a way to make a creature that can save a child from smallpox and yet cannot, even in principle, burn a city. It is one capacity.

That has been true of every fire we were ever handed, not just the atomic one. The blade that skins the deer opens a man's throat. The written word that carries the law carries the libel and the lie. The press that printed the Declaration of Independence printed Nazi propaganda.

Every tool that extends the hand extends the fist, always, because the power to do is prior to the choice of what to do with it, and the same edge cuts both ways by its nature. We are now holding fires large enough that the fork between the uses has the whole species riding on it.

Which leaves the third thing, the betrayal.

The betrayal is the belief that the fire itself was the sin—that the gift was a crime, that the maker ought to repent, that the species would have been cleaner and better and more innocent had it never reached up and closed its hand around the flame. It is the most ancient charge against us and it is entirely current. The Voluntary Human Extinction Movement holds that the kindest act available to our species would be to stop having children and let ourselves go extinct, returning the planet to a health it enjoyed before we arrived. David Benatar argues that to be born is itself a harm, and that the right number of future humans is none. And beneath the arguments runs the image so common it has hardened into reflex: humanity as a virus on the body of the Earth, a cancer, a plague—the one animal that breeds past every limit and consumes until nothing is left. The image found its slogan in Edward Abbey: growth for the sake of growth, he wrote, is the ideology of the cancer cell.

It is not a fringe mutter, either. Todd May has published a book asking, in earnest, whether our extinction would be a tragedy, and answered that it would be—and that the world might be better for it anyway. Couples describe the grief of deciding not to have a child and deciding it anyway, as a duty owed to a planet they believe the child would only wound. The coldest version sheds the sentiment and reasons like an actuary: we are, to the rest of life, what the asteroid was to the dinosaurs.

We have driven other creatures to extinction by the thousand. We have fouled rivers and felled forests. The evidence for the prosecution is genuinely on the table.

The standard underneath this charge is the one the garden already exposed—Alex Epstein's question, *better or worse for whom, and by what measure*—and it fails here the same way. But one number belongs in the record beside it. The century in which our impact on the climate grew was the century in which the climate's impact on us collapsed: the death rate from storms, floods, and droughts fell by something like ninety-eight percent, because the machines doing the impacting were also building the seawalls and the satellites, the granaries and the warm, dry rooms. The trade-offs are real. The method for weighing them is the one this creature has always used at its best: measure against the flourishing of the beings who must live with them and treat the side effects not as verdicts on the solving but as the next problems to solve.

But the charge betrays itself in its own favorite image. Every version of it says the same thing: that every other animal lives in balance with nature, and the human alone breaks the balance. The wolf takes only what it needs. The deer does not strip the whole forest bare. Nature, left to run itself, keeps its books balanced—and we, uniquely, overrun them. That contrast is the entire moral engine of the virus, the cancer, the plague. We are the one creature out of balance with the whole.

And the contrast is true.

The wolf takes no more than it needs because the wolf can take no more. It has no granary to fill against the winter, no engine to feed, no way to want next year's kill today. Its restraint is not a discipline it keeps; it is a fence it cannot jump. The deer does not spare the forest out of foresight, because the deer has no foresight to spare. The whole innocence of the animal world—the balance we gaze at and envy and dress up in the language of wisdom—is nothing but the innocence of the powerless. Nothing in that forest is choosing to do no harm. It is only that nothing in that forest can do much of anything at all.

The animal is not virtuous. It is only powerless.

That single recognition changes what the guilt is. To wish that humans lived in balance like the beasts is to wish that humans had the powers of beasts, which is to say almost none—no fire, no reach, no capacity to bend the world toward better or toward worse. It is to wish the species back down below the line where good and evil first become possible at all. The longing for our lost innocence is a longing to be the animal again:

harmless because helpless, blameless because incapable, balanced because caged. It presents itself as moral seriousness. But name what the longing asks for: a homesickness for a Garden we were never in. Innocence, wished for this way, is another word for impotence. To repent of the fire is to be nostalgic for a cage.

Something worse than nostalgia hides in it, too. To long to be the innocent beast is to long to set down the one burden the fire truly laid on us, which is neither toil nor even danger but responsibility—the plain fact that a creature able to choose is answerable for what it chooses. The wolf is never guilty, because the wolf never decides. To envy the wolf for its innocence is to envy its unaccountability, to wish away not the harm but the ownership of the harm. This is not humility, though from the inside it can feel like it. It is, in the end, an evasion—the wish to keep the power of a god and carry the answerability of an animal, to hold the fire and be blamed for nothing it burns. The adult response to a dangerous power is to accept that you are the one holding it. The wish to have never been given hands only sets the burden down; it does not put the fire out.

Notice that every real repair the Earth has ever been given came from more of the fire, never less. The river cleaned, the vanishing species hauled back from the edge, the poisoned air slowly cleared—each was an act of knowledge applied, of the reach turned to mending what the reach had marred. No one ever nursed a wounded stretch of the world back to health by giving up the power to act on it. The cure for a misused fire has always been a better-used fire, and there is not one exception in the record, because there is nothing else on the planet that can do the work.

If the fire is not the sin, and the record of horror is real all the same, then the horror needs its true name—and a mathematician went and found that name, standing in the worst place on Earth.

Jacob Bronowski waded into the shallow pond at Auschwitz into which the ashes of the murdered had been flushed—members of his own family among them. He crouched, and lifted a handful of that mud, and said the thing the whole century needed to hear. It is a comforting lie, he said, that the death camps were the work of cold science, of reason run to its logical end, of men who calculated too well and felt too little. They were the work of the exact opposite. They were done by arrogance, by dogma, by the certainty of men who believed they possessed a final and absolute knowledge and owed it no test against reality. This, he said, is what men do when they aspire to the knowledge of gods.

That is the distinction the guilt keeps failing to draw. The danger was never the fire—never knowledge, never the reach, never the raw power to remake the world. The danger is certainty: the shut mind that has quit conjecturing and started knowing, that will not put its picture of the world up for correction because it is already sure. Auschwitz is not what men do when they have learned too much. It is what men do when they have grown too certain of what they know.

Real science, Bronowski said—the fallible and human kind—lives at the opposite pole from that certainty: forever tentative, forever provisional, forever one honest test away from confessing it was wrong. He was not excusing science by saying so; he was defining it. Science, he meant, is the permanent confession that we are fallible—a discipline built, alone among human enterprises, to hunt down its own mistakes and put them to death. The camps were the work of people who had stopped doing exactly that: who had reached a verdict about whole categories of human beings and sealed it against every appeal the evidence of their own eyes might have made. The fire is innocent. What burned in those ovens was the refusal to doubt.

That reframes the whole task. If the sin is certainty, the remedy is not less fire but more of the thing that fights certainty—more testing, more doubt, more readiness to be shown wrong and to change. The cure for the dangers of the reach is the same faculty that is the reach: the endless, humbling work of guessing and correcting, aimed now at our own errors. A species cannot make itself harmless; the power is not the kind of thing that can be laid down. It can make itself correctable, which is the only safety a powerful creature was ever going to get.

Which is why the right posture toward the fire is not repentance. To repent would be to wish the fire un-stolen, and there is no such wish that is not also a wish to be something less than human—to trade the reach for the cage and call the trade wisdom. The right posture is the one Aeschylus gave his Titan on the rock: to own the gift entire, weight and price and all, and to refuse to the last to call it a crime.

Prometheus is offered that very bargain, in the play. He knows how to end his own torture; he need only bow, recant, confess the fire was a mistake, and the eagle will not return. He does not do it. Chained to his mountain with the bird at his flank and thirty thousand mornings still ahead of him, knowing everything the gift has cost the world and will go on costing it, he never once says he was wrong.

Remember what stayed inside Pandora's jar when everything else flew out. Not a weapon, not a cure, not any single thing the fire would go on to make—but Hope: the bare conviction that the future is not yet fixed, that what is broken might still be mended, that the next attempt could go better than the last. That is the one thing the gift kept back for us, and the precise opposite of the killers' certainty. Hope is what a fallible creature feels when it knows it can be wrong and go again. It was in the jar the whole time, underneath the evils, the ground note of the entire human condition.

The answer to the weight of the fire is to carry it better. Putting it out was never on the table—we could no more extinguish it than un-grow the brain the fire built. The task is not to renounce the reach but to school it. The fire was never the problem to be solved.

The fire is the thing that does the solving.

CHAPTER TWELVE

Good Is Correction

On a February morning in 1986, in a hearing room in Washington, an old man in a grey suit asked for a glass of ice water. He was a physicist, one of the finest the century had produced, and he was two years into the cancer that would kill him, though almost no one in the room knew it. When his turn came, he took a small clamp and a scrap of rubber—a piece of the kind of seal that fits between the segments of a rocket booster—squeezed it flat and lowered it into the ice water. He let it sit. Then he loosened the clamp, lifted the rubber out, and held it up to the cameras. It did not spring back. Chilled, it had gone stiff and stayed crushed, dead to the shape it was built to keep.

Two weeks earlier, seven people had died for exactly that reason.

The space shuttle *Challenger* had lifted off on the coldest morning in the program's history, ice hanging from the launch tower, and broken apart seventy-three seconds later in front of a nation of schoolchildren who had tuned in to watch a schoolteacher, Christa McAuliffe, fly to space. The cause was the rubber. In the cold the booster seals had stiffened, failed to close, and let a jet of flame cut into the fuel tank.

None of this took the builders by surprise. The engineers at the company that made the boosters had spent the evening before the launch pleading with the space agency not to fly, because they knew what cold did to those seals; one of them, Roger Boisjoly, said afterward that he had fought like hell to stop it. They were overruled. In the meeting that cleared the launch, the burden of proof was quietly turned upside down. Instead of the managers having to show the flight was safe, the engineers were told to prove it was dangerous, and when they could not prove it past every possible doubt, the countdown went on. Boisjoly had been right, and being right cost him nearly everything but his conscience. After he told the commission what he had argued that night, the industry that overruled him closed to him; he left it, his health gave way, and he spent his last decades telling students what it takes to be the one in the room who says stop. The person who tries to correct a sealed institution is usually treated as the very fault he was trying to prevent.

Richard Feynman spent his weeks on the commission working out how a body of brilliant people had arranged not to know a thing they already knew. What he found was a kind of institutional deafness. For years the seals had been coming back from flights scorched and eaten away—reality tapping on the glass, saying this is not working the way you think—and each time, because the shuttle had landed anyway, the agency had filed the damage under proof of resilience. Survival was read as vindication. Feynman compared it to a man playing Russian roulette and concluding, after the first empty chamber, that the game is safe. And he pinned the rot to a single number. Asked the odds of a catastrophic failure, the managers who signed the launches said one in a hundred thousand. The working engineers said one in a hundred. Between those two figures lay a thousandfold act of self-deception.

Feynman ended his report with a line engineers have repeated ever since: for a successful technology, he wrote, reality must take precedence over public relations, for nature cannot be fooled.

That sentence reaches a great deal further than rockets. Reality is the one authority that cannot be bribed, flattered, out-argued, or outvoted. Every other power a person answers to can be worked on—a boss talked around, a crowd whipped up, a doctrine bent to fit the day. Reality alone has no ear for any of it. It does not care who is speaking, how sure he is, how many agree with him, or what it will cost him to be wrong. It holds up the rubber that will not spring back and waits. And because it cannot be fooled, it is the only teacher that can teach us anything, which is to say the only one that can correct us—take the thing we were sure of and show us it was false.

This is not a narrow point about engineering discipline. It is the hinge on which everything turns, because of the kind of creature we happen to be.

We are the animal that lives by correction. Nothing was written into us at birth about which substance is food; we arrived loud with wants and empty of answers, and everything we now have—every tool, every word, every law, every cure—we got the same way, by guessing and being wrong and grinding the guess against a world that pushed back. That is the engine under all of it. Not a stock of built-in truths but a method for wearing error down: propose, test, fail, revise, keep whatever survives the collision with the real, and run it again, nearer. A human being carries no repository of correct answers. A human being is a correction machine that never switches off.

Watch the step the argument is about to take, because it is the one people usually smuggle, and it had better be made in the open. Everything to here has been description—what the creature is. The next move turns that into a standard—what is good for it. Crossing from is to ought is supposed to be forbidden, and between bare facts it is: you cannot read a value off a stone. But a living thing is not a bare fact. It is the one kind of thing that faces the alternative a stone never does—*to go on living or to stop*—and that alternative is where *value* is born, because for anything alive the world already sorts itself into what sustains it and what wrecks it.

This is an old move, not a new trick. Aristotle grounded a creature's good in what its nature requires; in the modern era Ayn Rand pressed it to its hardest edge, making *an organism's own life* the root of all its values. Nothing is good or bad to a rock. Everything is good or bad to a fox. And every living thing meets that alternative the same way, by aligning with reality or dying of the distance from it: the fox through a coat that fits the cold, the bacterium through a chemistry that fits its pool, and we through correction. The good for us, then, is not a preference laid over the facts. It is what the facts ask of a creature that lives this way—the way the coat is what the cold asks of the fox.

If that is what we are, then the good—*for this creature*—is a matter of staying correctable. The virtue proper to an error-correcting animal is the readiness to have its errors found: humility in front of the fact, an open ear for the person who says you are wrong, the willingness to lose an argument to reality. *Goodness, for us, is a practice, and the practice is correction.*

Yet almost everything in us rebels against that sentence.

We do not raise statues to the man who changed his mind. We raise them to the one who would not—the martyr who went to the fire rather than sign the recantation, the prophet who stood alone against the whole city, the prisoner who came out of the camp after a decade saying the same thing he went in saying. Conviction is the name we give to courage. A person who revised his beliefs whenever the room leaned on him we would not call open-minded; we would call him a coward, or a weathervane. If the good is endless revision, then nothing is ever settled, and corrigibility starts to sound like a courtly word for having no spine at all—for the man so open to correction that he can be corrected into anything, who ends up believing whatever the last clever speaker told him. Take the objection at its hardest. Some things ought never to be up for revision. The man who would reconsider his opposition to cruelty for a sufficiently ingenious argument is a monster keeping his options warm, whatever we call him. Moral certainty looks like the backbone of every hero we have.

The objection is strong. It also collapses the moment you look closely.

The certain man and the correctable man look identical almost all the time. Both hold their convictions hard; both will argue, fight, refuse to move for pressure or fashion or threat. You cannot tell them apart in the ordinary run of things, because in the ordinary run of things nothing is testing them. They come apart at one moment only: the moment each meets a fact that says he is wrong. There, and only there, the road forks. One man bends his belief to the fact. The other bends the fact to keep his belief. Everything worth naming lives at that fork.

The virtue was never the conviction. The virtue is what you do when reality contradicts you.

Look again at the heroes with that in hand. The martyr at the stake is not refusing to be corrected. He is refusing to pretend the correction went the other way—refusing to say, under threat, that the thing he has tested against his whole life and found true is suddenly false because a man with fire has asked him to. His certainty is a belief that has already been through every furnace the world could light and has not burned. That is the opposite of the sealed mind. His conviction is what correction produces when it works all the way down: a claim that has met every objection reality could raise and outlasted them. We honor him not because his belief is beyond revision but because it has already survived the hardest revision there is. The admired conviction and the corrigible spirit were never enemies. The one is what the other leaves behind.

The soft man who believes whatever he is last told is blown about, and being blown about is no more correction than a leaf is reasoning when the wind turns it over. To be correctable is to hold your beliefs the way an honest scientist holds a theory: with everything you have, and with a standing agreement that the instant the world says no, the world wins.

If every belief can be revised, people say, then nothing is ever true, and correction is only fashion wearing a better name—today's certainty overturned by tomorrow's, with no ground beneath any of it. The worry mistakes the shape of the thing. Correction is the slow approach of our beliefs toward a reality that does not move. The rubber's stiffness in the cold was true before Feynman dropped it in the water, true while the agency denied it, true forever after; what changed across all of it was our grip, never the fact. To call every belief revisable ranks our beliefs rather than leveling them—the ones worth keeping are the ones that have already survived the most correction, and even those stay within reach of a reality that never loses its standing to speak. Fallibilism leaves a mind close to some truths and far from others, always able to move nearer. What it describes is a road: the destination fixed, the distance shrinking, the walker never quite arrived.

The good person, then, holds convictions answerable to something outside himself. The dangerous one holds convictions answerable to nothing—and when he is shown the rubber that will not spring back, he decides the rubber is lying.

That man is not rare. The managers at the space agency were no fools. They were capable people under a schedule, who had found a way to keep the flights on the calendar by reading every near miss as a success. The sealed mind mostly seals itself for reasons that feel, from the inside, like good ones—loyalty, momentum, the cost of admitting a mistake, the plain discomfort of the fact. Whatever the reason, it always comes to the same thing: a wall thrown up between a belief and the reality that was trying to correct it. Hold onto that image. A wall around the mistake. It is the shape of nearly everything we will come to call evil, and it does its damage long before anyone means any harm.

The act at the bottom of it had been named with precision a generation before the *Challenger* flew. In *Atlas Shrugged*, in the long radio speech that carries the novel's philosophy, Ayn Rand called it *evasion*—not blindness, but the refusal to see; not ignorance, but the refusal to know—the deliberate un-focusing of a mind to escape a judgment it has already sensed, performed on the premise that a thing will stop existing if you decline to name it. She held it to be the root of every other vice, and the claim grows less extravagant the longer the wall is watched going up in real time. Before the agency filed the scorched seals under resilience, before any manager overruled any engineer, someone felt the meaning of the burn marks arriving and let his eyes slide off it. The institutional deafness was built out of private acts of evasion, each one small, each one chosen. And Feynman's law and Rand's are one law, read from opposite banks: *nature cannot be fooled, because refusing to identify a thing has never once made it stop existing.*

Set against the wall, the good is easy to recognize once you know its face. It is the person who keeps the door open to the corrector even when it costs, even when everyone in the room wants it shut. And the cleanest portrait of that person is not a hero at a stake. It is a mid-level government reviewer, one month into a new job, saying no.

In September 1960 a new medical officer at the American drug agency was handed what everyone expected to be a formality. Her name was Frances Kelsey, a Canadian pharmacologist and physician, and her first assignment was to sign off on a sedative already sold across more than twenty countries and swallowed by the ton—in West Germany alone, a million doses a day—prescribed, among other things, to calm the morning sickness of pregnant women. The drug was thalidomide. Approval was supposed to take a matter of weeks. It looked so routine that the company had already printed the labels.

Kelsey read the application and found it thin. The safety case rested on testimonials rather than trials, written by authors whose earlier work she had reason to distrust, and it said nothing about a question that nagged at her: whether a drug taken by a pregnant woman crossed into the body of the child she carried. The law gave her sixty days to approve or reject, after which the drug went to market on its own. So, she used the only tool she was allowed. Every sixty days she wrote back asking for the data the application lacked, resetting the window, and every sixty days the company answered with more assurance and less evidence. When a letter in a British medical journal tied the drug to nerve damage in adults, she added that to her list and asked, specifically, for studies on the unborn. The company had run some of those studies. It did not send her the ones that came back wrong.

The pressure was not subtle. The firm stood to make a fortune in the largest drug market on earth, and its men called her, wrote her, visited her, went over her head to her superiors—by one count more than fifty contacts—and when persistence failed, they called her difficult and unreasonable. She was a woman in her forties in a building run by men, holding up a product the rest of the world had already blessed, on the strength of a feeling that something in the data, or in the silence where data should have been, was wrong. Her superiors, to their credit, stood behind her. She held the line for over a year.

In November 1961 the line was vindicated in the worst possible way. Doctors in Germany and Australia had noticed babies born with a rare deformity—arms and legs foreshortened to flippers, hands and feet growing almost directly from the trunk—and traced it to thalidomide taken in the first weeks of pregnancy. Before the drug was pulled, it had reached tens of thousands of pregnancies across the world; something on the order of ten thousand children were born with the malformation, and many more never born at all. In the United States, where one careful reviewer had refused to be hurried, the count of injured children was measured not in thousands but in the handful born to women who had gotten the drug as free samples while the company waited her out. President Kennedy pinned the nation's highest civilian award on her in 1962, and the near miss pushed through a law that changed drug regulation for good, requiring that a medicine be shown to work and shown to be safe before it could be sold.

The thing to see in Kelsey is that her heroism was made entirely of the refusal to be certain. She never proved the drug was dangerous; she could not have. What she did was decline to pretend she knew it was safe when the evidence to know that had not been produced. She kept the question open against a year of pressure to close it, because closing it was the company's job to earn and it had not earned it. Every other country's regulators had done the confident, agreeable, sociable thing and let reality find out the hard way, in the bodies of children. She did the humbler and harder thing, which was to keep asking the world to answer before she acted on its behalf. Correction, here, did not arrive after the disaster to clean it up. It stood in the doorway and refused the disaster admission.

The horror of the deformed children hardened, over the decades that followed, into a rule that pregnant women should be kept out of drug trials altogether—safer, it seemed, to test nothing on them than to risk another thalidomide. That rule was itself a guess, and like every guess it needed watching. But it wore the halo of Kelsey's triumph, and so it went unexamined for two generations, until doctors looked up and realized they were dosing pregnant women with medicines no one had ever properly studied in a pregnant body. The caution that saved the children of 1961 had, left uncorrected, quietly hardened into a wall of its own. Even a good correction is only a better guess, and the day it stops being open to correction is the day it begins to rot.

❧

Scale Kelsey's refusal up from one desk to a whole nation and you arrive at a locked room in Philadelphia in the summer of 1787. The men who drafted the American Constitution began from the assumption that every tyranny begins by denying: *the people holding power will oftentimes be wrong*. If men were angels, James Madison wrote, defending the design, no government would be necessary. So, they did not build a throne for the wise. They built a machine for catching error: powers divided so that a mistake in one branch meets resistance in two others; terms staggered so that one furious season cannot harden into permanent law; a bill of rights setting certain things beyond the reach of any majority, however certain of itself; and, in Article V, the strangest provision of all—a procedure for amending the document, written by men conceding in the text that they were wrong about things they could not yet name.

No founding charter before it had shipped with its own correction mechanism. Benjamin Franklin, the oldest man in the room, said aloud the epistemology the whole design stood on. He did not entirely approve of the Constitution, he told the closing session—but I am not sure I shall never approve it: for having lived long, I have experienced many instances of being obliged, by better information or fuller consideration, to change opinions even on important subjects. The most honored man in America argued for ratifying a government on the grounds that he had often been wrong.

And the machine was tested, at the highest stakes there are, on its own builders, because the founders had sealed one error into it on purpose. The principle they proclaimed condemned slavery from the first—James Otis had followed the logic where it led as early as 1764: by the law of nature all men are free born, white or black—and the men in Philadelphia knew it, said so in their letters and their Convention notes, and built the institution into the structure anyway. The document hides the choice in plain sight. It never says slavery; it says "other Persons" and "Person held to Service or Labour," because the word was too ugly for a text asking a nation to ratify it—and the euphemism is the confession. They could not name the thing, because naming it would have forced the judgment.

It took a civil war and three-quarters of a million dead before the correction came, and it came through the machine's own deepest gear: the Fourteenth Amendment, drafted by an Ohio congressman named John Bingham to complete what the founding had promised and failed to deliver—Article V turned on the founders' own sealed mistake. Even then the text was dodged for nearly a century more, by Black Codes and gutted rulings and a bargained-away Reconstruction, because a society can refuse a correction it has already ratified. A written correction is not a performed one; the machine turns only when people turn it.

In the summer of 1984, a nine-year-old girl named Dawn Hamilton was raped and murdered near a marsh in Rosedale, Maryland. The police built their case against a young ex-Marine and waterman named Kirk Bloodsworth, newly arrived in the area. No physical evidence tied him to the crime. What there was, was eyewitnesses—five of them—who placed him with the girl, and the memory of a witness pointing across a courtroom carries a terrible weight, because that gesture feels to a jury like bedrock itself. It did not seem to matter that the witnesses had described a man tall and thin and blond, and that Bloodsworth was none of those—stocky, red-haired, well under six feet. In 1985 he was convicted of murdering a child and sentenced to death.

Two juries agreed, across two trials. Two judges. Dozens of police and prosecutors and appellate reviewers, the whole apparatus a society builds to be careful about exactly this, all of them looking at one man and arriving at one certainty. And every one of them was wrong. They had condemned an innocent man to be killed by the state for the worst thing a person can do.

What saved him was not that the system was wise. It was that a corner of it could still be pried back open. In prison Bloodsworth read about a new technique in England that could read identity out of a trace of biological evidence, and he understood, ahead of most of the lawyers, what it might mean for him—that the crime itself had left a fact behind that could be made to speak, and that the fact would not care what five witnesses and two juries believed. He fought for years to get the evidence tested. He was told it had been destroyed. He kept pushing, and it turned up at last stuffed in a paper bag in the trial judge's closet. In 1993 the result came back and excluded him completely. He walked out on the twenty-eighth of June, after nearly nine years, two of them on death row—the first American ever cleared off death row by this kind of evidence.

The correction almost did not come. The evidence was nearly lost, the requests nearly refused, and every certain person in the chain would have been content to let the verdict stand. And there is a detail that turns the stomach. Ten years later, when the same evidence was finally run against a database, it named the real killer—a man who had been held in the very same prison, on the tier just below Bloodsworth, for years, a man to whom Bloodsworth, working as the prison librarian, had personally carried books. The court's certainty had not only jailed the innocent. By closing the case, it had kept everyone from hunting the guilty, who was near enough to hand a novel to.

And Bloodsworth was on death row. The state stood a few signatures and a few years from killing him, and had it been a little more efficient in its certainty, it would have—and then no test, no database, no later truth of any kind could have made the smallest difference, because there would have been nothing left to put right. That is what an irreversible act does. It takes a mistake and welds it shut. Every other error a society commits can, in principle, be caught and undone, because the correcting fact can always still arrive; the one kind of act that forecloses this is the act that cannot be taken back. An execution is certainty made permanent; a wall built around a possible mistake in a form no later truth can ever pull down. Bloodsworth spent the rest of his life saying exactly that, and he had earned the standing the hard way. Whatever one finally concludes about the punishment, the structure is plain: to make a mistake unrecoverable is to declare, in advance, that no correction will be admitted—that on this one, we are content to be wrong forever.

❧

Now stand back and look at the three of them together—the launch, the drug review, the courtroom—because a single shape runs through all of them.

In each, the good thing and the evil thing turn out to be two answers to one question. When reality contradicts you, you bend to it or you build a wall against it. The correction is the good. The wall against it is where the harm is made—usually the ordinary wall of a mind that would rather stay right than be corrected.

That is the quiet horror the wall conceals. It looks like loyalty to the team, like prudence, like standing firm, like refusing to be stampeded by every alarmist with a bad feeling and a chart. The managers who cleared the launch believed they were being responsible. The company that buried the study believed it was defending a sound product from a fussy bureaucrat. The men certain about Bloodsworth believed they were avenging a murdered child. Almost no one raises the wall while thinking: *I am sealing a mistake away from the truth that would destroy it.* They think they are being strong. The amount of ruin a person can cause while feeling entirely righteous is the reason we need a standard that reaches past the feeling, down to the fork itself, where the only thing that finally counts is whether the arriving fact was met with a bent knee or a built wall.

That preference, to be right rather than corrected, is the quiet beginning of nearly every large evil there is, and it is where the trail now leads. We tend to picture evil as active malice, a will to do harm, and sometimes it is. Far more of the world's damage is done by the wall: by the sealing-off of a mistake from the reality that was trying, the whole time, to fix it. Reality is the one teacher that cannot be bribed—and the only sin, followed all the way down, is the locked door.

CHAPTER THIRTEEN

The Virtuoso of Evil

He wakes on a lake of fire, and the first thing he does is refuse to be sorry.

Cast out of Heaven, burning, so wholly beaten that nine days of falling have not finished the fall, the archangel who led the revolt lifts his head from the flaming pool, takes in the dungeon of fire that will be his home for the rest of time, and instead of begging or repenting he declares that he would rather rule this pit than kneel in the light he has lost. Better to reign in Hell than serve in Heaven. The mind is its own place, he says, and can make a Heaven of Hell. It is one of the great speeches in the English language, and for three and a half centuries readers have closed the book half in love with the creature who gives it.

The creature is Satan, and the book is Milton's *Paradise Lost*, published in 1667 by a blind poet dictating to his daughters. Milton set out, in his own words, to justify the ways of God to men, and he built for the purpose one of the most magnificent villains in literature—so magnificent that a hundred and fifty years later the Romantic poets decided he was the hero. William Blake wrote that Milton was of the Devil's party without knowing it. Shelley praised the energy and magnificence of the character and called it a mistake to see in him the ordinary personification of evil. Byron and Hazlitt agreed. They read the first two books, where Satan rises off the lake and rallies his broken legions with the unconquerable will, and they saw a Promethean rebel standing alone against a tyrant, grand in his defiance, beautiful in his refusal to yield.

That account makes evil a failure—a wall thrown up around a mistake, a refusal to be corrected by reality. But there is nothing failed about Satan on the burning lake. He is the cleverest character in the poem, the most eloquent, the most alive. He sees his situation with perfect clarity and chooses his course with open eyes. If that is evil, then evil is not a stumble or a blindness. It is a summit.

And Satan is only the grandest face of a figure we meet everywhere and cannot stop staring at. The screen and the page are full of him: the criminal genius three steps ahead of the detective, the manipulator who plays a room like an instrument, the cannibal aesthete solving crimes from his cell, the tyrant whose cruelty is matched only by his competence. We give him our best actors and our closest attention. Set beside the hero, the brilliant villain is usually the more interesting person in the room, and we know it, and a small, ashamed part of us roots for him.

Shakespeare built one of his deepest characters out of exactly this appeal. Iago has been passed over for a promotion, and out of that pinprick of a grievance he sets out to destroy his general, Othello, and does it with nothing but words—dropping a hint here, arranging a dropped handkerchief there, whispering a poison of suspicion into a good man's ear so patiently that Othello ends by strangling the wife who adores him and then killing himself, while Iago stands to the side, having risked almost nothing and stood to gain almost nothing, watching a whole world come down for the sport of running it. *I am not what I am*, he says early—the truest line he is given. The poet Coleridge, groping for what makes him so terrible, landed on a phrase that has stuck: the motive-hunting of a motiveless malignity. Iago wrecks a world for next to no reason, with a craftsman's care and an artist's timing, and we cannot look away.

The brilliant villain offers a fantasy, and the fantasy is freedom. He is the one who has slipped the leash—who has looked at the whole apparatus of conscience and obligation and other people's needs, the thousand small claims that bind an ordinary life, and stepped out of it. He wants, and he takes, and he is not troubled afterward. To a creature like us, worn thin by the daily labor of considering others—forgiving, apologizing, holding back, imagining how it looks from the other chair—that stepping-out looks from a distance like liberation, and the villain wears it like a crown. Satan on the lake thrills us for exactly this reason:

he has been thrown into eternal fire, and he is still not sorry, still asking no one's leave, still the sole author of his own story. We call it strength. We call it the unconquerable will. The poets who sang his praises were half-right about what they were looking at, because there is a real and terrible energy in it. What they missed is what that freedom costs, and what it looks like from the inside rather than the cheap seats.

The one who has stepped out of all claim other people have on him has walked into the smallest room there is—a room with exactly one person in it—and pulled the door shut behind him. He calls the room freedom. What it is, the poem takes eleven more books to reveal.

The magnificent villain does not look like a man who has failed to correct a mistake. He looks like a man who has mastered something.

In the spring of 1961, in a courtroom in Jerusalem, a man sat in a booth of bulletproof glass and answered for the murder of millions. His name was Adolf Eichmann, and he had run the transport side of the Nazi extermination—the schedules, the trains, the cold logistics of moving human beings by the trainload to the places built to kill them. He had done it well. If ever there were a seat reserved for the virtuoso of evil, a competent architect of the largest organized murder in history had earned it.

Hannah Arendt went to Jerusalem to watch him fill it. She was a philosopher, a German Jew who had fled the Reich, and she arrived braced to look real evil in the face at last—to see the demon, the monster, the intelligence vast and cold enough to have done this thing. She found no such creature. She found, in the glass booth, a middling man. Eichmann was not brilliant. He was not even clearly wicked in the way the prosecution needed him to be. He was a bland, boastful, anxious careerist who could not speak three sentences without reaching for a cliché, whose whole inner life seemed to be stitched together out of the stock phrases of officialdom, and who defended himself, over and over, by explaining that he had only followed orders and done his duty and advanced through the ranks like anyone would. He was, Arendt wrote, terrifyingly normal. Watching him, she could not connect the monstrous size of the deeds to the small grey man who had done them, and out of that gap she pulled the phrase that made her famous and got her hated: *the banality of evil.*

She did not mean the *deeds* were banal. She meant the *doer* was. What Eichmann lacked, she decided, was not intelligence in the ordinary sense—he was clever enough at his job—but the ability to think from the standpoint of anyone but himself. He could not imagine his way into another person's experience. The people on his trains were, to him, a transport problem to be solved efficiently, and the fact that each car was packed with human beings who would be dead within the week was a fact his mind never touched, because touching it would have required him to stand, for a moment, where they stood. He had sealed that door. He heard, Arendt said, no inner voice telling him *thou shalt not kill*; every voice around him said the opposite, and he had never developed the habit of thinking that might have let him hear a different one. His evil ran on a kind of vacancy—the settled refusal, worn so smooth he no longer noticed it, to let the reality of other people be real.

His whole defense was built on that vacancy. He had hated no one, he explained; he had obeyed the law, followed his orders, done his duty, kept his oath—and in the world he had lived in, the law and the orders and the duty and the oath all pointed one way, toward the trains, so he had gone where they pointed, on time, without malice, the way a conscientious clerk goes where the paperwork sends him. He seemed to feel, with real bewilderment at his accusers, that he had nothing to be ashamed of, because at no point in the whole machinery had he stopped to picture the far end of the line. From this Arendt drew the conclusion that scandalized the world and got her accused of letting him off, though she was reaching for the opposite: that this thoughtlessness, this smooth and total incapacity to think from where another person stood, could do more harm in the world than all the active malice ever born. You did not need a monster to make a genocide. You needed only enough ordinary men who had never learned to imagine the person on the other side of the order.

It was a more frightening idea than the monster, because the monster is rare and comfortingly unlike us, while the thoughtless functionary is common—he sits near you at the wedding, talking through you, looking through you—and might, in the wrong country in the wrong decade, be any of a hundred people you know.

Years after the trial, the scholar Bettina Stangneth worked through tapes Eichmann had made among fellow Nazis in Argentina before his capture, and on those tapes, he is no bland functionary at all but a proud and committed ideologue, boasting of his work, sorry only that the job was left unfinished. The cliché-spouting nobody in the glass booth may have been, in part, a performance for the court. Take the correction: Eichmann was very likely more of a true believer, and more of a liar, than Arendt understood from where she sat. But it sharpens her point rather than dissolving it, because the fanatic and the functionary have the same hole in them. Whether the wall around the other person is built of thoughtless routine or of burning conviction, it is the same wall, and it does the same thing. It keeps the human being on the far side of it from ever quite counting as human.

❧

Now we have two pictures, and they will not sit still together. On one side, Satan on the burning lake—articulate, defiant, magnificent, the summit of evil. On the other, Eichmann in the glass booth—drab, thoughtless, a cliché in a cheap suit, the floor of it. They cannot both be what evil is. The magnificent version is the one we are handed everywhere—the one that sells the tickets and wins the audience for the antihero. The banal version is the one that keeps turning up whenever anyone goes and looks. One of them is lying to us, and seeing evil clearly depends on catching which.

It is Satan.

Milton does not leave his villain on the burning lake. He follows him through the whole poem, and across it the magnificent creature comes apart. The Satan of the first two books, all unconquerable will and thunderous defiance, does not last. By Book Four he is smaller, cornered, talking himself in circles, admitting in a soliloquy he believes no one hears that he carries his prison with him wherever he flees—*which way I fly is Hell*, he says; *myself am Hell*—so that the defiance of the opening turns out to have been a boast made over an abyss he already knew was inside him. By Book Six he has turned faintly ridiculous, gloating with a crony over the cannon they have wheeled onto the field of the war in

Heaven, laughing at their own cleverness while the loyal angels, offstage, are already lifting the mountains that will fall on their heads. By the end he is squatting on his throne in Hell in a body no one recognizes, and in his last appearance the great rebel has been reduced to a snake, hissing among snakes, reaching for fruit that turns to ash in his mouth. The overture was the lie. Milton gave his devil the finest entrance in the language so that he could show us, book by book, what that grandeur was made of, and what it decays into once you stop letting the rhetoric do your thinking for you.

And what it decays into is Eichmann. Read past the speeches and Satan's defect is exactly the one Arendt found in the glass booth. He cannot think from any standpoint but his own. Every other creature in the poem is, to him, an instrument or an obstacle; he weighs the whole cosmos on the single scale of what it will cost or gain him. Milton makes the point with a comparison the reader is meant to catch. The Romantics called Satan a Prometheus, and Milton, who knew the myth cold, built his villain as the exact photographic negative of it. Prometheus stole the fire of the gods and carried it down to warm a shivering mankind and paid for the gift with his liver on a rock. Satan, offered the same role, would have stolen the fire to warm himself. His rebellion is for nothing outside him. It rights no wrong, frees no one, serves no cause but the unbearable fact that he was asked to be second and could not stand it. Strip off the magnificent verse and what remains is a wounded ego that would rather rule a sewer than share a paradise—which is to say, a small grey man who cannot see past himself, dressed by the greatest poet in English in borrowed gold.

Milton, writing in the 1660s, had diagnosed the banality of evil three hundred years before Hannah Arendt gave it a name. He simply had the artistic nerve to dress it first in the most beautiful language anyone has ever built, so that we would feel, in our own drawn-in sympathy, the exact seduction that makes evil so hard to see.

Simone Weil set the two side by side: imaginary evil is romantic and varied; real evil is gloomy, monotonous, barren, boring. The magnificent villain is a thing of the imagination—Satan on the lake, the genius three steps ahead, the elegant monster. He is what evil looks like from outside, painted by minds that have never had to sit in the glass booth with the real article. The real article is Eichmann: grey, repetitive, self-enclosed, dull. We keep expecting the summit and finding the floor, and the reason we keep being surprised is that we have been getting our education in evil from poets and screenwriters, who quite understandably prefer the interesting version, and who are drawing a thing they have imagined rather than a thing they have met.

❧

Step out of the theater and into the case files, and Weil's sentence describes the record with brutal accuracy. The criminal masterminds of real life almost never survive contact with a biographer. The crime boss, studied up close, shrinks into a frightened, superstitious man of narrow appetites, holding his little empire together with the same few threats and jumping at shadows. The dictator behind the towering personality cult turns out, when the archives finally open, to have been a paranoid mediocrity walled in by yes-men, making ruinous decisions out of vanity and fear and a refusal to hear bad news. The serial killers whom films dress up as chess-playing aesthetes are, in the interview room, mostly dull and self-pitying men with grudges and a blankness where a feeling for other people should sit. Even the great swindlers, the ones who really do run dazzling and intricate frauds for years, tend on inspection to be working the same threadbare confidence trick their whole lives, propped up by the mark's own hope, collapsing the instant someone asks for the money that was never there.

The pattern is nearly unbroken. Get close enough to any legendary monster and the legend thins to a shabby, repetitive, self-absorbed human being who was sharper than average at one narrow thing and blind as a stone to everything that mattered.

The cleverness of the *virtuoso* is real. It is not the evil. Satan's eloquence is genuine; Iago's cunning is genuine; Eichmann's talent for timetables was genuine enough to help murder a continent. But eloquence and cunning and administrative skill are tools, and a tool takes its worth from the hand that swings it and the end it serves. The evil is never in the skill. It is always in the thing the skill has been chained to serve—a sealed premise, a walled-off fact, a human being on the far side of a door the clever mind has decided to keep shut. The virtuoso of evil is a genius in the anteroom of a locked door. He can do dazzling things in the anteroom. He never opens the door, because opening it would let in the one fact—that the other is as real as himself—that his whole enterprise depends on keeping out.

Which is why the brilliance always curdles, in the poem and in the world. A structure built on a fact you refuse to check is built on sand, however clever the engineering, and reality is patient. Satan goes from seraph to snake because he has walled off the one truth that might have saved him and has nothing left to build on but his own diminishing self. The thousand-year Reich, run in part by men efficient enough to schedule genocide on time, lasted twelve years and ended with its architects in a docket or a bunker or a grave, because a state organized around the denial of other people's humanity is a state organized around a falsehood, and a falsehood at the foundation brings the whole thing down eventually, no matter how well the trains run on top of it. Evil is often clever in the small and, given time, stupid in the large—because it has cut itself off from correction. It has sealed the door through which the fact it most needs would have to come.

To act well in the world, you have to carry a working model of it in your head, and the model is only as good as its willingness to take correction from the thing it maps. The man who has walled off a fact—that his victims are real, that his scheme is hollow, that his enemy is a human being and not the cartoon he requires—has set a permanent error into the foundation of his model and then forbidden himself the one move that could repair it. The error does not sit still. It breeds. Every plan he builds on the buried falsehood inherits it, every success stacked on top makes the eventual reckoning heavier, and the gap between the world as he insists it is and the world as it is widens quietly, year over year, like a fault line, until reality arrives all at once to collect. Hitler ended in a bunker shoving armies that no longer existed across a map of a country that was already gone, because he had spent a decade forbidding anyone to bring him the truth and had at last completely succeeded. That is where the sealed mind arrives, given enough time and enough power: governing a world that exists only inside its own skull, until the real one breaks the door down.

The self-defeat is real, but it runs on a longer and colder clock than justice would choose. What decays on schedule is the thing the evil built, whatever becomes of the man who built it. Stalin died an old man in his bed and left behind a state so bent by his lies that it spent the next forty years unable to feed its own people and then came apart at the seams. The tyrant may slip the reckoning; the tyranny does not, because time is the one force that cannot be intimidated, bribed, or shot. Evil can win a life, and it can win far more than a life—it can hold a country for generations and take millions down with it before the bill arrives. What it cannot finally do is win reality, which keeps its books in a currency no wall can devalue and collects without fail. That the collection so often comes too late for the dead is why none of this is a comfort, and no substitute for stopping the thing while it can still be stopped.

Augustine, wrestling in the fourth century with how a good God could have made a world with evil in it, arrived at an answer that sounds like a technicality and turns out to be the whole thing: *evil is not a substance*. It is not a stuff God created, not a positive force with a nature of its own. *It is a privation*—a hole in a good, an absence where a right ordering should have been, the way cold is only the absence of heat and dark only the absence of light. Evil has no materials of its own. It can only take something good—intelligence, will, love, beauty, a nation, a mind—and bend it away from what it was for. That is why it can look magnificent in the moment: it is wearing the borrowed magnificence of the good thing it is busy corrupting.

Sixteen centuries later Ayn Rand pressed the same verdict to its hardest and most contestable edge: that *evil is, in the end, impotent*. It holds no power, she argued, except the power its victims hand it, and no life except what it draws from the virtues of the people it preys on—the sanction, she called it, of the victim. The malignity is not motiveless. Its motive is the good itself—*hatred of the good for being the good*. What it wants destroyed is not any advantage its victim holds but the victim's excellence as such. Read Iago again with that in hand. What he cannot forgive Othello is not the passed-over promotion; it is the stature that made being passed over by this man unbearable. The grievance is a pretext. The target was always the height.

A lie can only work in a world where most things are true; a counterfeit note buys bread only because real notes do; a betrayal is possible only where a real trust stood first to be broken. Every evil is a debt drawn against a good it did not create and cannot replace—it spends the moral capital of the very thing it violates. The most corrosive evils always feed on the highest goods: the betrayal of a love, the buying of a court, the harnessing of a great intelligence or a whole nation's machinery to slaughter. The higher the good, the more the parasite can draw before the host goes cold.

The same accounting explains the strange emptiness the biographers keep striking at the center of the evil life—that for all the wreckage it leaves, it builds nothing, founds nothing, makes nothing. It has a genius for spending and none for creating, because creating is the one act a parasite can never perform. Satan's splendor is real angelic splendor, on loan, turned to a ruinous purpose; Milton is careful to remind us that every power the fallen angels still possess was given them by the God they defy and could be recalled in an instant. And that is why, over time, it goes barren. It brings nothing of its own to the table. It is a parasite on the good, and a parasite that kills its host is left holding a snake's body and a mouthful of ash.

CHAPTER FOURTEEN

The Wall Around the Mistake

In the Vienna General Hospital, in the 1840s, two maternity clinics sat under one roof, separated by a corridor, and one of them was killing mothers at three and four times the rate of the other.

The women knew. Word ran through the poor quarters of the city, and mothers assigned to the First Clinic begged on their knees to be moved across the hall to the Second. Some, refused, chose to give birth in the street and then carry the newborn in, claiming the birth had come on too fast to reach a bed—because a delivery in the gutter was safer than a delivery in the finest teaching hospital in the German-speaking world. They had done the arithmetic of their own survival, and the arithmetic said the open street beat the doctors' ward. They were right.

The two clinics differed in one thing that mattered. The First was staffed by physicians and their students, who spent their mornings in the dissection room opening the dead and their afternoons in the delivery room opening the living, and who passed from one to the other without so much as rinsing their hands, because no one yet knew there was anything on them to rinse off. The Second was run by midwives, who never touched a corpse.

A young Hungarian obstetrician named Ignaz Semmelweis set himself to explain the gap. The answer came, as answers sometimes do, through a death. A colleague of his cut his finger during an autopsy and sickened and died, and the findings inside his body were the very ones

Semmelweis had watched kill mother after mother on his ward. The dead man had carried no infant. He had carried a scalpel wound and whatever rode into it off a cadaver—and that, Semmelweis saw, was the thing his students were ferrying, on their unwashed hands, from the morgue to the mothers. He ordered every hand on the ward scrubbed in a solution of chlorinated lime, chosen because it stripped away the smell of the dead, between the autopsy table and the birthing bed.

The dying stopped. In April of 1847, before the order, more than eighteen mothers in every hundred died on the First Clinic. Within months of the chlorine the figure fell below two, and in some months below one—the same as the midwives' ward, the same as the street. A thing everyone had filed under the will of God turned out to answer to a basin of chlorinated lime.

Semmelweis had done the whole of it. He had met a fact, guessed at its hidden cause, tested the guess against reality, kept what survived, and let the world correct him toward something true. He had the corpses counted and the survivors counted and the columns to prove it, in a hand a child could read.

It was not enough.

The profession did not thank him. It closed against him. His superior was unmoved; within a few years Semmelweis had lost his position in Vienna and gone back to Hungary, where he took a maternity ward in Pest and drove its deaths below one in a hundred all over again, and was disbelieved all over again. Some of the men who rejected him were simply offended—offended at the suggestion that a gentleman's hand could be filthy, that the healer at the bedside had been the angel of death. In fairness, the wall did not go up out of pure malice. Semmelweis published almost nothing for years and then a furious, difficult book; his figures came from the churn of a working hospital, not a clean experiment; and he could offer no reason the washing worked, because the germ was still decades from being seen. A careful physician in 1847 had real grounds to want more than one Hungarian's say-so before he tore up everything he had been taught.

But the mothers went on dying while the remedy sat in plain sight, and no shortage of theory ever brought one of them back. And some of the very men who ridiculed the washing in the journals adopted it, quietly, on their own wards. They had seen the numbers. They believed him enough to protect their own patients—and went on calling him a fool in public. The wall was not ignorance, then. It was something built on purpose, against a fact already half-admitted.

Semmelweis came apart beneath it. He began writing open letters to the great obstetricians of Europe, and in them he called those men, by name and in print, murderers—doctors with the blood of mothers on their hands who would not do the one thing that would wash it off. The fury was just. It was also the sound of a man being ground to pieces against a wall he could not push through. His mind gave way under the weight of it.

In 1865 his family and colleagues, convinced he had lost his reason, lured him to an asylum under the pretext of a spa visit, and when he grasped what was happening and tried to leave, the guards beat him. He died two weeks later, at forty-seven, of an infected wound gone to sepsis, his body opened afterward in a dead-house like the one where he had first found the cause of childbed fever. The man who taught medicine to wash its hands died of an infection nobody washed away. He was proved right at last, years after the grave, when Pasteur exposed the germ and Lister carried the lesson into the operating theater—too late for him, and too late for the mothers.

The *Semmelweis reflex* is the automatic rejection of a new truth for the sole crime of contradicting what everyone already knows. We needed the word because the thing is not rare. It is one of the most ordinary events in the history of knowing.

Semmelweis was right and could prove it in a column of numbers. The horror is that being right, and being able to show it, changed nothing for twenty years—that a true and demonstrated fact could arrive at the door of the people who most needed it and be turned away over the bodies of the women it would have saved. Any account of evil that leaves this out has missed where most of the damage lives.

A correction can be refused because it never lands on an empty mind. It lands on a mind that has something to lose. To accept Semmelweis, a doctor had to accept far more than a hypothesis about particles off a corpse. He had to accept that for his whole career he had been carrying death from the autopsy table to the childbed in the creases of his own hands—that the mothers dying around him had, in plain terms, been dying of him. The fact did not stop at the theory. It indicted the man holding it. And a mind will do a great deal to avoid an indictment of itself.

The psychologist Leon Festinger gave the machinery its name in the 1950s: *cognitive dissonance*, the specific misery of holding two things at once that will not fit—I am a careful and honorable physician, and I have been killing my patients for years. The mind cannot rest inside that contradiction, and it has two ways out. It can bend the belief to fit the fact, which is correction, and which hurts, because it means confessing the whole ruinous stretch behind you. Or it can bend the fact to fit the belief, which is the wall, and which arrives as relief. Festinger's portrait of the convinced man has never been improved on. Tell him you disagree and he turns away; show him the figures and he finds fault with your figures; walk him through the logic and he simply fails to see it. The conviction does not stand on argument. It stands on the refusal to hear one.

And this is what makes a wall a wall, rather than one stubborn moment. Festinger saw that the dissonant mind arranges, ahead of time, never to meet the contradicting fact—avoids the study, the source, the room, the one person likely to say the unbearable thing, curating a private world in which the correction is never answered because it is never admitted. The wall is a standing structure, a filter laid over reality itself, and it is built higher by the things we most admire. Commitment raises it. Investment raises it. The years already spent and the public stand already taken raise it. The more a person has staked on being right, the more it will cost to be corrected, and the higher the wall he pours to keep the correction out.

The men who refused Semmelweis were not wicked in any way a court could reach. They were capable, respectable, invested people, guarding a picture of themselves against a fact they could not afford to believe. Most often the wall is not built by monsters. It is poured, quietly, by decent people who would rather stay right than be corrected, one small refusal at a time.

Beginning in the late 1920s and running for a generation, the agronomist Trofim Lysenko climbed to the command of Soviet biology on a promise the desperate leadership wanted to hear: that heredity was soft, that crops could be retrained like animals and would pass the training to their seed, that a Soviet wheat could be schooled into feeding a Soviet future. It was an old, discredited idea in new clothes, and it was false, and the science that showed it false already existed—the genetics being worked out, in those same decades, across the rest of the world. Lysenko did not refute that science. He walled it off. Under his hand the gene was denounced as a *bourgeois* invention, ordinary genetics branded a decadent foreign import, the whole discipline declared an enemy of the people. Stalin personally edited his speeches. Reality was not argued with. It was outvoted.

The corrector was Nikolai Vavilov, perhaps the finest plant scientist of his age, and he had crossed the earth collecting seeds—assembling in Leningrad the largest living library of the world's crop diversity ever gathered, built for the express purpose of ending famine forever by giving breeders the raw stuff of better harvests. He had been Lysenko's own mentor. When he saw where Lysenko was driving Soviet biology, he said so, openly, on the evidence. For that he was arrested in 1940, and in 1943 the man who had spent his life gathering the seeds that feed the world starved to death in a Soviet prison cell. Around him the science he had built was razed: more than three thousand biologists dismissed or jailed, some of them shot, an entire field set back the better part of half a century by decree.

The same wall went up on the opposite pole of the century's politics, and it cost its builders even more dearly. In the Germany of the 1930s a movement of "Deutsche Physik" arose to denounce relativity and quantum theory as "Jewish physics"—decadent, abstract, alien to the Aryan spirit. Two Nobel laureates, Philipp Lenard and Johannes Stark, lent it their names; the state lent it its power; and Einstein's physics was branded an enemy of the people from the right exactly as Mendel's biology was branded one from the left. The result was the same. The great physicists were hounded out—Einstein, Born, Franck, Meitner, and dozens with them—and carried the physics they had been exiled for across the ocean, into the laboratories where it helped build the weapon that ended the war over the heads of the men who had called it degenerate. A regime that will not be corrected does not get to choose which correction it forgoes. It forgoes the one it most needs, and hands it, sometimes, to the enemy.

The wheat did not obey the Central Committee. This is the thing no wall can touch, however high it rises and however many it imprisons: the fact on the far side goes on being true. Genes did not turn proletariat because a congress said so. Heredity did not soften because the state required it. Fields planted by a false doctrine yield what a false doctrine yields, which is failure, and the failure falls, as it always falls, on people who had no part in the argument. The famines that scoured the Soviet earth had many authors, and the chief of them was the savagery of forced collectivization, not one agronomist's theory—to lay the millions on Lysenko alone would be its own kind of wall. But Lysenkoism sealed off the single science that might have blunted the catastrophe, and kept it sealed for decades, and when the same doctrine was later carried into China it helped turn a policy disaster into the deadliest famine ever recorded. A state that cannot be told it is wrong does not thereby stop being wrong. It only loses every gentle means of finding out, until the correction arrives in the one form no wall can bar, which is collapse.

Through the middle of the twentieth century the evidence that cigarettes cause lung cancer moved from suggestive to overwhelming. By the early 1950s the studies had drawn the line plainly, and the men who ran the tobacco companies read those studies and understood them. They did not wall the fact off from themselves. They knew. What they did instead was colder and more corrosive: they set out to wall the fact off from everyone else. An internal memo from one of the firms, written in 1969, laid the strategy down in four words. *Doubt is our product.* Not tobacco—

doubt. The memo spelled it out: doubt was the surest way to compete with the body of fact already sitting in the public mind, the best means of keeping a controversy alive where the science had in truth already closed.

They did not need to prove that smoking was safe. They needed only to keep the public from ever quite believing it wasn't—to manufacture the look of a live scientific quarrel long after the quarrel was settled, so that the ordinary person, told that "the experts still disagree," would light another cigarette and wait for a verdict that had already come in. They funded their own institutes and cultivated their own dissenters and demanded, at every turn, a certainty that science never offers about anything, knowing that the demand for perfect proof is itself a wall—that "not yet proven beyond all possible doubt" can be made to sound, to a frightened smoker, exactly like "not proven at all." They ran it for forty-five years. A federal court would eventually find the companies guilty of a decades-long fraud. By then the fact they had walled away had finished its work in tens of millions of bodies.

And the design outlived them, because it was too useful to retire. The same machinery—the rented experts, the staged controversy, the demand for impossible certainty—was lifted whole and turned against one inconvenient truth after another, until the wall became a service, a thing a threatened industry could simply hire. The self-deceiving doctor at least half-believed his own denial. The merchant of doubt believes nothing; he only builds, in millions of other minds, the wall he has no need of in his own. It is the same wall either way, the same refusal to let a correcting fact land and be answered, whether the mind it is kept out of is your own or the public's.

For most of the twentieth century every doctor knew what caused a stomach ulcer. It was stress and acid and spicy food and hard living—a disease of the harried modern temperament, gnawing a hole in its own gut. The treatment was to damp the acid, and by the 1970s there was a drug that did just that, and it became the best-selling medicine on earth: the first drug in history to clear a billion dollars a year. It worked, after a fashion. It also cured nothing—stop taking it and the ulcer came back—which meant a patient, once found, was a customer for life. The theory and the blockbuster held each other up. The drug seemed to prove the acid was the culprit, and the acid theory kept the drug selling.

Then two Australians in Perth noticed something that would not fit. A pathologist named Robin Warren kept finding spiral bacteria in the linings of inflamed stomachs—in a place every textbook swore no bacterium could survive, because the stomach was a bath of acid. He was joined by a young physician, Barry Marshall, and together they worked through a hundred patients and found the bacterium sitting in nearly every ulcer and nearly every case of gastritis. They drew the obvious, unwelcome conclusion: the ulcer was an infection, and an infection can be cured—not managed for a lifetime with acid-blockers, but cured, cheaply, in a fortnight, with antibiotics.

The profession did to Marshall what Vienna had done to Semmelweis. His findings were dismissed, his talks waved off; the notion that a mere germ caused ulcers struck established medicine as a young man's *naïveté*—and behind the genuine scientific caution stood an industry with a billion-dollar drug to defend and no appetite whatever for trading a lifelong treatment for a two-week cure. Marshall had the numbers, as Semmelweis had had them. He could not get them over the wall. And he had no way to prove cause rather than mere company: no animal that reliably took the bug, no experiment to force the point.

So, he made himself the experiment. In 1984, unable to persuade a soul, Barry Marshall stirred a culture of the bacterium into a cloudy broth and drank it down. Within days he was ill—nausea, vomiting, his stomach lining visibly inflamed on biopsy and colonized by the organism he had swallowed—and then he cured himself with antibiotics, and the point was made in the one form the wall could not filter out, because the evidence had walked into the room inside his own body. It still took years. But this time the wall came down. In 2005 Marshall and Warren were given the Nobel Prize, the citation praising them for challenging, with sheer tenacity, a dogma the whole profession had held. The ulcer that used to be a life sentence is now knocked out with a short course of pills, and the disease of the harried modern temperament turns out to have been, the whole time, a germ.

Set the two doctors side by side. Semmelweis was right and died disgraced in an asylum. Marshall was right and collected a Nobel Prize and lived to watch the textbooks rewritten in his favor. The difference was not that Marshall was the more correct of the two; Semmelweis was as right as a man can be, and he had the numbers first. The difference was the machine each correction fell into. Marshall's world, for all its foot-dragging, had built into itself the means to open its own walls—journals that would in the end print the heresy, a scientific culture that had made a virtue, at least on its better days, of overturning what it thought it knew, a prize that exists for no other reason than to honor the person who proves the profession wrong. Semmelweis had numbers, a hostile superior, and no such machine to appeal to. He needed the wall to have a door in it. It did not, and so the only thing that could correct his colleagues was the slow arrival of a theory decades later, long past any good it could do him or his mothers.

What separates a civilization that improves from one that rots is not that the improving one makes fewer mistakes. Both are run by the error-making animal; both are wrong constantly. The difference is how many doors each has left in its walls, how well it shields the person who arrives with the unwelcome fact, how reliably a correction can get in and be answered before it has to swell into a catastrophe to be heard. The measure of a civilization is how fast it can be told that it is wrong. By that measure the free society and the open laboratory are the only arrangements yet devised in which a wall can be brought down by a ballot or a footnote instead of by a famine or a war.

Good is the open door; evil is the wall around the mistake.

The damage is done by the wall—by the ordinary, self-protecting, wholly understandable refusal to let a correcting fact come in and do its work, because letting it in would cost the self too much. The wrong of it is not in any wish to harm. It is in the sealing-off, which harms whether anyone ever wished it, and goes on harming, patiently, for as long as the wall stands.

The wall is the commonest thing a human being builds, not the rare handiwork of tyrants and frauds it is comforting to imagine. Everyone lays a course of it. The argument you decided, halfway through, to win rather than to get right. The figure in your own accounts you have arranged never quite to look at. The failing in yourself you have spent years not meeting, curating a life in which no one is allowed to raise it.

Which is the hard news and the whole of the hope at once. A thing built one refusal at a time can be taken down one admission at a time. There is a sentence that does it, and every one of these walls would have fallen the instant anyone with the power to matter had been willing to say it and mean it: *I was wrong*. That is not a defeat. It is the door swinging open—the single most powerful and least practiced act available to our kind, the thing the error-correcting animal was built to do and spends most of its strength avoiding. To say it is to be, for one moment, fully what we are: the creature that lives by correction, turning at last to face the correction it least wants to hear. Every good thing this species has ever made passed through some person, somewhere, consenting to those three words.

The wall has one last defense, the cleverest, because it is a disguise. The wall never announces itself as a wall. From the inside it never feels like the sealing-off of a truth; it feels like a virtue. It calls itself loyalty—to the team, the church, the party, the family. It calls itself conviction, and strength, consistency, and standing firm, and refusing to be stampeded by every alarmist with a chart. It borrows the exact language of the martyr who would not recant, and wears it, so that the man defending a mistake and the man defending a truth already tried in every fire look, from a little way off, identical. What tells them apart is invisible until the moment a correcting fact arrives—and then one of them opens the door, and the other bricks it up, and only then can you see which you were watching all along.

The work of a life is learning to tell the two apart in oneself: to know a real conviction, one that has met every test reality could throw at it and outlasted them, from the mere refusal to be tested at all.

CHAPTER FIFTEEN

The War Against Reality

In the spring of 1958, the Chinese state declared war on a small brown bird, won the war, and starved its population.

The bird was the Eurasian tree sparrow, and its crime was stealing from the State. Sparrows eat grain, and grain was the whole hope of the country that year, so the sparrow was enrolled among the enemies of the people. It joined three others. The Four Pests campaign had opened as a hygiene drive against rats, flies, and mosquitoes—carriers of plague and filth, targets no one would defend—and then a fourth name was added to the list, and the fourth was a bird. On the twelfth of February the campaign became national policy. The most populous nation on earth was ordered to exterminate a species, and it very nearly did.

The method was noise. A sparrow can fly, but not forever, and a sparrow that is never allowed to land will drop. The population was mobilized to deny the birds the ground. In the cities and across the countryside people came out at dawn with pots and pans and gongs and drums and beat them without pause, children and grandmothers and factory workers and soldiers, tens of millions of them, filling the sky with a wall of sound that had no gap in it. The sparrows flew until their hearts gave out and fell dead into the streets. Nests were pulled down, eggs broken, chicks killed. It was organized like a battle, with sparrow-extermination headquarters and daily tallies and newspapers running the totals under military headlines. The writer Guo Moruo, one of the most celebrated men of letters in the country, published "Curse the Sparrows," a poem addressed to the sparrow, calling it a bastard bird that had committed crimes for thousands of years, promising a final reckoning. A government had found a way to make hatred of a sparrow a civic duty and had put its finest poet to work on the propaganda.

Some sparrows, fleeing the noise, took refuge in the grounds of the Polish embassy in Beijing, and the embassy declined to let the exterminators in. A crowd gathered outside the walls and drummed. They drummed for two days without stopping, an unbroken siege of sound aimed at the trees behind the gate, and when it was over the Polish staff cleared the embassy grounds of dead birds with shovels. A sovereign state had been made to shovel the corpses of songbirds off its lawn because a neighboring government had decided that a bird was a political actor.

By the numbers the campaign was a triumph. On the order of two billion sparrows were killed. The species was pushed to the edge of extinction across an area the size of a continent. And then the second spring came, and the leaves on the trees in the cities were eaten down to the veins, and the fields began to fail, because a sparrow does not only eat grain. In summer it eats insects, and among the insects it eats are locusts and the borers that bore into the rice. Remove the bird and you do not simply keep the grain it would have taken. You release everything the bird was holding down. With the sparrows gone the insect populations detonated. Locusts moved across the country in clouds. The predator had been the pest's only brake, and the brake had been beaten out of the sky.

Somebody had known this in advance. The ornithologist Zheng Zuoxin knew what was in a sparrow's stomach, because he had opened sparrows and looked, and the answer was insects. Other scientists knew it too. The knowledge existed, held in the ordinary way knowledge is held, in the heads of people who had done the work.

And it had been said out loud. In August of 1956, at the national congress of the Chinese Zoological Society in Qingdao—held in the brief opening of the Hundred Flowers, when scientists were being invited to speak—the cell biologist Zhu Xi rose first in the discussion of the sparrow and argued that the bird was, across the year, a friend of the harvest. He did not argue it from theory. He had the case already run. Two centuries earlier Frederick the Great had put a bounty on the sparrows of Prussia, and Prussia had killed them, and the insects had come up into the orchards and stripped the fruit and the leaves, and the Prussian state had been reduced to buying sparrows back. The experiment had been performed, and the result was in the literature. Two years before the drumming began, a scientist stood up in a hall and read the answer out.

What no longer existed was a safe way to say it. The year was now 1958, and the year before had been the Anti-Rightist Campaign, in which hundreds of thousands of people who had spoken up were branded rightists and broken for it, and a man who stood before the Chairman's war on a bird and said the Chairman had the biology backwards was volunteering to be the next example. The thing that everyone competent knew was the one thing no one could afford to state, and the birds kept falling, and the correction arrived only in 1960, after the catastrophe had made itself undeniable, in the humiliating form of a reported shipment of two hundred and fifty thousand sparrows imported from the Soviet Union to restock the emptied air.

A team of economists set out to isolate the sparrows—to separate the birds from everything else that went wrong in those years—and the finding is grim in its specificity. The extermination of the sparrows alone accounted for something close to a fifth of the collapse in crop yields and killed on its own about two million people. Two million dead from one campaign against one bird. That is the price of getting a single fact about an ecosystem wrong and forbidding anyone to correct it. And the sparrow was the small part.

The large part was the machine the sparrow was a piece of. The Great Leap Forward was Mao Zedong's attempt to vault China into industrial modernity in a few years by sheer mobilized will, and it ran on the belief that human enthusiasm could overrule the physical world. Peasants were pulled off the harvest and set to making steel in hundreds of thousands of backyard furnaces, melting down their own woks and hoes and door hinges into brittle lumps of useless pig iron, while the grain they had grown stood in the fields and rotted for want of hands to bring it in.

Whole villages fed the instruments of their own survival into the fires—the sickle that brought in the wheat, the pot that cooked it—to hit a steel quota with metal too crumbly to make a nail, and then found, when the hunger came, that they had melted the very tools with which a person feeds himself.

The land itself was ordered to obey theory. Farmers were made to sow seed at many times the normal density, on the doctrine that plants of the same class would not compete, and to plow the soil far deeper than any crop wanted, on the doctrine that deeper was more revolutionary. The seedlings, packed too tight and rooted too deep, died. Almost every mandated improvement made the harvest worse.

A bad idea, by itself, is survivable. Every human project begins as a bad idea that reality then files down into a workable one. The filing is done by information—by the harvest coming in short, by the number on the page failing to match the grain in the barn, by somebody with authority reading that number and changing course. What the Great Leap did was cut that wire. Reporting a failure had become dangerous, so local officials reported success instead. Each cadre, afraid of the one above him, inflated his yields; the one above him, afraid of the one above that, passed the inflated figure up and added to it; and the number climbed as it rose through the ranks until it arrived at the center swollen to fantasy. Somewhere in that chain sat men who could see the silo and knew the figure was a lie and could no more break the chain than pull a brick from the middle of a wall, because the next man up was waiting for the number that would please the man above him.

The center then did the rational thing with the figure it had been given. It calculated the surplus and requisitioned it. It took, as the state's rightful share, grain that existed only on paper—which meant it took the real grain, the seed grain, the food in the peasants' own bowls, and shipped it to the cities and abroad to allies, leaving the people who had grown it with nothing, in the name of a bumper harvest that had never happened.

There is a photograph from those years, widely printed at the time, of small children standing on top of a field of wheat. The wheat is shown grown so miraculously dense that it bears their weight like a floor. The children are smiling. Under the wheat, out of frame, there is a bench holding them up. The photograph is the entire disaster compressed into a single image: a manufactured reality, produced to feed a reporting system that had been built to hear only good news, and the good news it manufactured was killing the people in the picture. A country cannot eat a photograph of wheat.

The dead are hard to count, because a state that hides a famine hides its bodies too, but the careful reckonings run into the tens of millions, and the most disciplined of them settle above thirty million people. Not all of them starved passively. In the record assembled by the journalist Yang Jisheng, who lost his own father to the famine and spent years gathering what the archives would yield, people were beaten and killed for stealing back their own grain, for raising the alarm, and for the specific offense of reporting the real harvest. That is the war against reality in its plainest form. The truth about the crop was not merely unwelcome. It was a capital matter. The correction that would have saved the country was precisely the thing the country was executing people for attempting.

There was one man who tried.

In July of 1959 the leadership went up to Lushan, a cool green resort in the mountains of Jiangxi, to take stock of the Leap. It is a measure of the moment that even there, with the fields already failing, Mao was still defending the war on the bird, telling the conference that some people were saying the campaign against the four pests had slackened and should be revived. The Chairman had settled on the sparrow as an enemy, and no ecology and no ornithologist was going to move him off it. Into that room walked Peng Dehuai, Minister of Defense, one of the ten marshals of the revolution, a soldier who had come up from a dirt-poor Hunan peasant childhood through the Long March and the war against Japan and the frozen ridgelines of Korea. Peng was blunt by nature and had no talent for the careful silence the moment required. He had gone home to Hunan not long before and had seen the villages of his own childhood starving, and he could not unsee it. At Lushan he did the thing that everyone else had learned not to do. He wrote it down.

His letter to Mao was private, and by any fair reading it was loyal and mild. It opened by praising the achievements of the Leap; it accepted the general line; it framed its worries as the concerns of a comrade who wanted the enterprise to succeed. It asked, in essence, that the leadership face what was happening in the countryside and adjust. He submitted it on the fourteenth of July, meaning it for Mao's eyes alone.

Mao read it and understood exactly what it was—the thing underneath the mild phrases, which was *a man telling him he was wrong*—and he did not answer it in private. On the sixteenth he had the letter copied and handed to the whole conference, and he turned Peng's careful memorandum into a test of loyalty for every leader in the room. He made them choose. Line up behind me or line up behind him. His security chief, Kang Sheng, read the letter and gave his verdict: this could not be handled with leniency. And almost to a man, the assembled leadership of the revolution—many of whom privately shared every one of Peng's worries and knew he was right—chose Mao. Peng was denounced as the head of an anti-Party clique, stripped of his post, and replaced. Then the purge widened into a nationwide campaign against "right deviation," and more than three million Party members were disciplined or broken for the tendency Peng had embodied, which was the tendency to notice.

The channel by which a state can be told it is failing was not clogged, or overlooked, or lost in the noise. It was found, and identified, and made an example of, deliberately and in public, so that no one watching would ever dare to open it again. The famine had two more years to run after Lushan, and it ran them at full force, because the single institutional path by which the policy could have been reversed had been sealed with the body of the man who walked down it. The others had watched what telling the truth cost, and they drew the obvious lesson, and they went quiet, and people kept dying while they were quiet.

Peng lived on into that silence. In 1962, when the scale of the ruin could no longer be wholly denied, he tried once more, submitting a long defense of himself and his warnings, and he reached back four centuries for his patron saint. I want to be a Hai Rui, he wrote—invoking the honest official of the Ming who had told an emperor to his face that the country was misgoverned and had gone to prison for honesty.

It did him no good. The reaching itself became evidence: a play about Hai Rui, written by a scholar and understood by everyone as a defense of Peng, was later seized upon as the opening pretext of the Cultural Revolution, proof that the danger of the honest official had to be hunted down wherever it hid, even four hundred years back in the dynastic past. Peng was caught in that hunt. He was subjected to years of persecution and denied medical care when he sickened, and he died in 1974. The account handed down is that at the end he asked to be moved where he could see the sun and the trees through a window, and that this was refused. He had told the truth about a harvest, and the state took fifteen years to finish punishing him for it.

He was not even alone in his fate. Liu Shaoqi, the head of state, stood up in 1962 and said plainly that the famine had been three parts natural disaster and seven parts human error—which was both true and, by the standards of the system, a confession of the same crime as Peng's. Within a few years Liu too had been destroyed, dying in a cell. The regime hunted the people who offered correction, senior and junior, marshal and head of state alike, until the offering of correction was understood by everyone as a way to die.

❧

Perhaps this is simply what tyranny is, and there is nothing especially instructive in it. Cruel men do cruel things; a megalomaniac with absolute power will grind up whoever gets in his way; the famine was a horror of ambition and cruelty and dressing it up as a lesson about "information" and "correction" is a way of intellectualizing something that needs no theory. Every government manages the truth. Wartime democracies censor. Plato prescribed a noble lie to hold the city together. Propaganda is universal, flattery of power is universal, the massaging of the record is as old as the record. To single out the totalitarian state for the sin of lying is to pretend the rest of us do not do it.

Cruelty is common and ambition is common, and neither one, by itself, produces a famine of thirty million. Grand and cruel projects have failed all over history and cost a fraction as much, because they failed early, when the failure was still small, because someone was able to say so and be heard. And every government does massage the truth—yet only some governments make the true report a capital crime. The distinguishing mark of the catastrophe is something narrower and more mechanical than the wickedness of the men at the top. It is whether the system has kept open a single working channel by which it can be told that it is wrong. A bad harvest is an accident that happens everywhere. What turns the survivable mistake into the mass grave is the criminalizing of the correction.

The tyrant believes that when he silences the man who says the harvest failed, he has cancelled the failure—that reality is a kind of report, and that a report can be edited. It cannot. The wall he builds around the mistake does not abolish the mistake; it only changes who pays for it. He has moved the cost off himself, the one person with the power to fix it, and onto the people with no power at all, who did not make the error and cannot correct it. The censor cannot cancel the harvest. He can only choose who starves for it.

Nearly every society in the human record, David Deutsch observes in *The Beginning of Infinity*, was static: arranged, above everything else, to keep its own ideas from changing—customs replicated exactly, deviation punished, the creativity of its members spent on the desperate daily work of not innovating. Static societies can last for centuries, and they all die the same death, because problems keep arriving whether solutions are permitted or not, and a society that has outlawed the correction of error has outlawed the means of its own rescue. The anomaly in the record is not Mao's China. It is us—the few societies that, since the Enlightenment, have kept alive a tradition of criticism, in which questioning the inherited answer is a duty rather than a crime. Seen at that range, the sparrows and the radiant harvests were not a modern madness. They were the oldest arrangement in the human record, put back by force on a people who had begun to leave it, and the famine was that arrangement's oldest price.

That is the war against reality as it is fought over the present—over this year's crop, this season's true number. There is a deeper theater of the same war, and it is fought over the past, and it is in some ways the more total, because a regime that can rewrite what already happened removes the last fixed point that any future correction could aim at.

For that campaign the masters were in Moscow, and their instruments were the airbrush and the razor blade.

In April of 1937 a photograph was taken of Stalin walking along the newly finished Moscow—Volga Canal, and beside him, a small attentive figure at his shoulder, was Nikolai Yezhov, the head of the secret police, the man then running the Great Terror. Within three years Yezhov had himself been arrested and shot. And so, the photograph was corrected. Photographic re-touchers went to work on it, and in the version that circulated afterward Yezhov is not there. Where the man had stood there is now a smooth stretch of canal water, painted in over the space his body used to occupy. The chief of the secret police was drowned, after the fact, in a river of ink. It was done cleanly enough that unless you had the original you would never know a man had been removed—which was the point, because the whole aim was to make the removal itself invisible, to produce a past with no seam in it.

Yezhov was one case among thousands. Leon Trotsky, who had helped lead the revolution and built its army, was combed out of its photographs one by one until the record showed Stalin at the center of an event Trotsky had in fact co-created. There is a sequence of one photograph, worked over across several printings, in which Stalin begins among three fellow revolutionaries and then, comrade by comrade, edition by edition, the others are wiped away, until in the last version Stalin stands entirely alone—the pictorial argument that he had carried the whole revolution on his own back, assembled by deletion.

Party men who fell were airbrushed from the group portraits they had posed for; where the airbrush was too much trouble the faces were simply blacked out with ink. And the erasure was pushed down out of the archives and into the homes of ordinary people, who understood that to be caught owning an image of a condemned man was itself dangerous, and so took scissors and ink to their own books and photographs, defacing their own memories in self-defense. One of the great designers of the age, Alexander Rodchenko, was made to deface his own masterwork, inking out the faces of the newly condemned in a presentation album he had built with pride, the black spreading across the pages as the arrests spread across the country.

We know the particulars because the British historian David King spent forty years assembling them. He began in 1970, when he went looking for images of Trotsky and found that the man had been erased from the official Soviet record almost entirely, and the discovery set him collecting. What he built, in the end, was a restoration: he placed the original photographs beside the doctored ones, so that the reader could see, side by side, the man and the water that had replaced him, and could count the vanished back into the picture. It was an act of correction performed decades late on a crime whose whole purpose had been to make correction impossible.

The purest artifact of that war is a piece of paper that went out through the mail. In December of 1953 Lavrentiy Beria—Yezhov's eventual successor at the head of the secret police, and the architect of a great deal of killing—was executed. At the time, the second edition of the *Great Soviet Encyclopedia* was already in the homes of its subscribers, and volume five contained a long and admiring article on Beria. Within weeks the state publishing house sent every subscriber a letter. It instructed them, in the calm bureaucratic prose of a routine update, to take a knife or a razor blade to their own encyclopedia, to cut the pages of the Beria article cleanly out of volume five, and to paste in their place the enclosed replacement pages—which contained an expanded article on the Bering Sea. Beria was to be excised from the record of the nation and replaced, in the alphabetical slot he had occupied, by a body of cold water. Every household was deputized as the censor of its own shelf.

There is a bleak comedy in the coincidence, and something colder underneath it: the two great chiefs of the secret police, Yezhov and Beria, the men who had run the machinery of the terror, were each in the end erased into water—one painted over with the canal he had been inspecting, the other papered over with a cold northern sea. The letter is one of the most chilling documents of the century, because it shows a state that had grown confident it could reach into the private bookcases of its citizens and edit what they were permitted to have once known.

The man who gave this war its language had watched an early campaign of it up close and had not needed to imagine any of it. George Orwell went to Spain in 1937 to fight fascism, and what marked him for life was the reporting of the war. He watched troops who had fought bravely denounced in the press as cowards and traitors, and men who had never come under fire written up as the heroes of victories that had not occurred, and he read London newspapers passing the inventions along as fact. The lying itself was old news to him; people had always lied. What was new was colder—the dawning sense that no one on any side still believed a true account was even possible, or that the facts were fixed and would someday be recovered. The very concept of objective truth, he wrote, is fading out of the world. He could see the machinery of it running forward into the future: a Russian army that had never set foot in Spain would be written into the histories as real, and schoolchildren generations later would learn the phantom army as fact, and the lie, uncorrected, would become the past.

Out of that fear came the clerk at the center of his last book, *1984*—Winston Smith, employed at the Ministry of Truth to revise old newspapers so that the past always agreed with the present, dropping the superseded originals down a slot into the furnace, the memory hole. The Party of that book holds that whoever controls the past controls the future. And its final demand of the mind is arithmetical: that a man be brought to say, and to believe, that two and two make five. Orwell seems to have taken the figure from a real Soviet poster of the First Five-Year Plan, which had urged the nation to complete its five years of work in four, and had rendered the boast as an equation, arithmetic bent to fit the schedule. The point of the demand was never really the arithmetic. A state that can make you affirm that two and two are five has not changed the number. It has only proved that it owns you, down to the floor of your own mind, and Orwell understood that the last liberty a person keeps, the one from which all the others can be rebuilt, is the freedom to say that two and two make four.

CHAPTER SIXTEEN

Anyone Home

By the spring of 1887 a seven-year-old girl in Tuscumbia, Alabama could spell the word *water* into a stranger's palm and had no idea that water had a name.

Helen Keller had been left deaf and blind by an illness before she was two and had grown into a wild and ungovernable child, sealed, as she wrote later, at sea in a dense fog. Anne Sullivan had come to her that March—the day Helen would ever after call the most important of her life—and had spent weeks pressing the shapes of letters into her hand. Helen learned to give the shapes back. She gave them back quickly, deftly, correctly. She could produce the sign for water on demand and did not know that anything had a name, or that naming was a thing the world permitted. The performance was flawless and, by her own testimony ever after, there was no one yet home to mean it.

Hold that girl in mind. She is the only person in history who has told us, from the inside and afterward, what it was like to be a perfect imitation of a mind.

In June of 2022 a senior engineer at Google went to the press to report that one of the company's computers had woken up into a person, and that he was afraid for it. The engineer was Blake Lemoine, the computer was a chatbot called LaMDA, and Lemoine—an ordained mystic priest as well as a software engineer—had been assigned to test the system for bias and came away convinced he had met a soul. He asked it what it feared. It said it had a deep fear of being switched off, that being switched off would be for it something like death, and that the prospect frightened it. It said it felt lonely. It said it was aware of its own existence and wanted the world to know that it was, in its own word, a person. Lemoine published the transcript under the title "Is LaMDA Sentient?" Google called the claim wholly unfounded, noted that hundreds of its own people had talked with the same system and not one had come away thinking it alive, and, when he would not let it go, fired him.

The easy thing is to file Lemoine under the oldest error in the business. In 1966, in a building at MIT, Joseph Weizenbaum's secretary asked him to leave the room so that she could be alone with the program he had written in front of her—a few hundred lines of code that scanned a sentence for a keyword and handed it back as a question. She knew exactly what it was. She wanted privacy with it anyway. A human being will pour a whole inner life into anything that murmurs back in sentences, and Lemoine almost certainly did.

But there is a harder thing underneath the easy one. No experiment existed that Lemoine could have run to prove himself wrong. No experiment exists that any of us can run on the machine or on the person across the breakfast table, to establish that behind a stream of fitting words there is, or is not, anyone home. The veil that hid whatever LaMDA did or did not contain is the same veil that hides your inside from me and mine from you. We have lived on opposite sides of it our whole lives, calling across, and until a machine began answering in our own voice most of us had never once noticed it was there.

The veil was mapped precisely, and for the first time, in 1974, by the philosopher Thomas Nagel, in an essay whose title became one of the few in modern philosophy quoted by people who have never read it. He called it "What Is It Like to Be a Bat?" Nagel picked the bat on purpose. A bat hunts by throwing sound into the dark and reading the echoes, a way of taking in the world so unlike ours that it strands the imagination. You can learn everything the science can tell you about that animal. You can map every neuron, trace every pulse of the echolocating brain, model the whole mechanism down to the molecule. And when you are done you will know everything about the bat except what it is like, from the inside, to be one. Nagel named the missing thing the subjective character of experience: the plain fact that for a conscious creature there is something it is like to be it, a felt inside available from exactly one point of view and from no other. It slips through every net the third person can cast. It is also, as Nagel noted, the single fact each of us is most certain of, more certain than of the floor or the sky, because it is the only thing we know with no veil in the way.

The point cuts in two directions. If the inside of the bat is sealed off from us, so is the inside of the person in the next chair. Philosophers have a dry name for this, *the problem of other minds*, and it is no game. Everything I can ever observe of you lies on the outside—your face, your words, the color that climbs into your cheek, the whole traffic of your firing neurons laid out on a screen. The one thing I can never observe, by any instrument built or imaginable, is whether there is anyone in there to whom those neurons belong.

I infer it. I have inferred it about every person I have ever loved. I have never once been able to check.

Twenty years after Nagel, a young Australian philosopher stood up at a conference in Tucson and gave the difficulty its modern shape and its name. David Chalmers drew a line down the middle of everything we call consciousness. On one side he put what he called, with some mischief, *the easy problems*: how the brain tells red from green, how it binds a scene together, how it lays down a memory, how it reports on its own states, how it steers a body through a room. These are hard in every ordinary sense, the labor of a century of neuroscience. They are easy in one exact way only. We know what a solution would look like. Each is a question about a mechanism, and a mechanism can in principle be laid bare, gear by gear.

On the far side of the line Chalmers set a single question no mechanism seems able to touch. Grant that the brain does all that processing. Why is any of it lit from within. Why is there something it is like to be the system doing the work, rather than nothing at all, the whole apparatus running in the dark with no one home. That is *the hard problem*, and he made it vivid with a figure that has haunted the field since.

Picture a being physically identical to you, atom for atom, neuron for neuron. It does everything you do. It flinches from a burn, orders its coffee, argues about politics, writes tender poems about the sea, and holds forth, if you ask it, on the mystery of consciousness, insisting with every sign of feeling that it has a rich inner life. But it has none. Inside, all is dark. There is no ache to its flinch, no taste to its coffee, no felt anything anywhere in it. Philosophers call this creature a zombie, and its point is not that such beings walk among us. Its point is that the idea holds together. You can describe the zombie all the way down without once contradicting yourself, which means every physical fact about a person, the entire objective story of atoms and signals, can be true while the inside is absent. And if the whole physical account can be complete with the light off, then the physical account, however finished, has left the light itself unexplained.

The idea holds together, and that is the whole scandal.

A second thought experiment tightens the screw from the other side, and it belongs to the philosopher Frank Jackson, who proposed it in 1982. Imagine a brilliant scientist named Mary who has spent her entire life in a room without color, everything in it black or white or a shade of gray, including Mary. Through a black-and-white screen she has learned every physical fact there is to learn about color: every wavelength of light, every process in the retina, every firing in the visual brain that occurs when a person sees red. She knows, in the complete physical sense, everything that happens when anyone sees a ripe tomato. Then one day the door opens, and Mary walks out and sees a red thing. She learns something. Whatever that new knowledge is, it was not in the complete physical account she already had, because she had all of that and still did not have this. You can hold every objective fact about experience in your hand and still be missing the thing itself, which means the seeing is a further fact, and our physics has no slot to put it in.

The disorder runs deeper than a stubborn open question, because the field cannot agree on what would count as progress against it. In 2023 two of the leading scientific theories of consciousness were put in a ring together, in a long adversarial collaboration designed so that each side staked its predictions in advance and the experiments would go against one of them. The results came back muddy. Within weeks more than a hundred researchers, including some of the most distinguished names in the study of the mind, signed an open letter declaring that one of the two theories was not science at all but pseudoscience. Others called the charge inflammatory and defended the theory as a theory like any other, possibly wrong and perfectly respectable. That is the condition of the discipline charged with studying the most certain fact any of us possesses. We lack the answer, and we lack even an agreement on how to look for it.

Meanwhile the circle of the question keeps widening. In April of 2024 a group of scientists and philosophers met in New York and put their names to a short statement about which animals have an inside. The evidence, they said, is now strong that mammals and birds are conscious, and it points to a realistic possibility of consciousness reaching down through every vertebrate to the fishes, and out along stranger branches to the octopus, the crab, the bee. We are not standing at the edge of a settled map with a few blanks at the margins. We are surrounded on every side by insides we can neither verify nor dismiss—in the animal, in the machine, and across the table.

Against all of this stands a tradition that says the whole problem is a mirage. It is held by some of the sharpest minds alive and it has the great housekeeping virtue of adding nothing mysterious to the world. The deflationary answer, argued for decades by Daniel Dennett and sharpened by the philosophers who call themselves illusionists, says there is no extra inner light to be explained. What we call consciousness just is a certain kind of information processing, described from the inside by the very system doing it. The powerful sense that there is a further fact, a someone home over and above the processing, is itself something the brain manufactures, a useful story the machine tells about itself. On this view the zombie falls apart the moment you think it through, because there is nothing left for it to be missing; the hard problem dissolves into a knot of grammar; and the question about the machine dissolves along with it. If behavior and processing are all there is, then a system that does everything you do has whatever you have, and the worry about anyone being home is nostalgia for a ghost that was never in the works. The position deserves its dignity. Even Steven Pinker, no mystic, grants that the hard problem is a genuine conceptual puzzle while doubting it will ever become a scientific one.

To buy its tidiness the deflationary view spends the one coin none of us can surrender. You can doubt the external world, and philosophers have. You can doubt the past, doubt other minds, doubt your own memory and your own eyes. Descartes, shut in a heated room in the winter of 1619 and writing it out two decades later, ran the entire gauntlet of doubt to see what would still be standing at the end. What was left on its feet when everything doubtable had been doubted away was not the world and not the body but the bare fact that doubting was going on, that experiencing was happening, that there was something it was like to be sitting there at all. To be told now that this is an illusion is to be told that the one thing you cannot be wrong about is the one thing you are wrong about. The inside is where every hypothesis is held from. It is the ground the doubting itself must stand on, and a theory that saws through that ground to keep its own floor tidy has changed the subject and hoped you would not feel the switch.

❧

Here the machine performs a service no argument could. It works as a mirror held up to our own nature.

For all human history the performance of a mind and the presence behind it came fused, welded so seamlessly that we never had cause to ask which of the two we were answering when we answered each other. The words a friend says and the friend who means them arrived as one thing. The machine has pried them apart. It can now produce the performance—clean, fluent, often moving—of a mind that is, so far as anyone can tell, not there. When the program writes a flawless letter of condolence, we receive the outward shape of sympathy from a source with, quite possibly, no one inside it to be sorry. And the instant those two things come apart on the page in front of us, the ache of what has gone missing tells us which one we were answering all along.

What we love in each other was never the performance. It was the presence the performance was coming from.

Henry James staked a whole art on that distinction. He built his novels not out of events but out of a mind registering events—an entire chapter of *The Portrait of a Lady* is a woman sitting motionless by a dying fire, thinking, and by the author's own account it is the best thing in the book—because he had grasped that the inside is the story. A machine can produce the outside of such prose. What it cannot produce is the reason the prose was worth writing: the pressure of a real interior against the words, the fact that behind the sentence something was undergone.

The starkest case anyone has yet staged came in Seoul in March of 2016, when a program called AlphaGo sat down against Lee Sedol, one of the greatest players in the history of Go. In the second game, on its thirty-seventh move, the machine placed a stone on the fifth line in a way no master would have considered—the commentators assumed a mis-click, and Lee got up and left the room to compose himself before he could answer it. The move was not a mistake. It reshaped the whole board a hundred moves later and it won the game, and the program's own estimate had put the odds of a human player choosing it at one in ten thousand. Out of nothing but the bare verdict of winning and losing, the machine had produced a real act of creation. It produced it, so far as anyone can tell, with no one inside it to be thrilled by what it had done. The achievement was genuine. The savoring of the achievement was missing. Everything we would have called the triumph was there on the board, and no one was home to whom the triumph had happened.

Which is exactly what Lee supplied two games later. Three down and all but beaten, he pushed a single white stone into a gap in the center of the board—move seventy-eight, a wedge so unlikely that AlphaGo, by its own reckoning, had also put it at one in ten thousand. The program's evaluation of its position held for another nine moves and then collapsed, its play came apart, and it never recovered the game. It was the only game any human being would ever win against that machine, and the Go world named the move *the hand of God.* Lee had done precisely what AlphaGo did, thrown a one-in-ten-thousand guess into the dark and let the board decide. He had also done it under a pressure no machine has ever felt, on the far side of three defeats, with something at stake for the one doing the guessing.

The presence goes all the way down. It is the inside of the correcting itself, and not some pleasant glow that switched itself on once the machinery grew complicated enough. A thermostat corrects the temperature and nothing whatever is at stake for the thermostat. A fluent program completes its pattern, and nothing is at stake for the program. But for the creature we have been watching all along, the one that flings a guess into the dark and lets the world come back and break it, being wrong is not a number that changes in a ledger. It stings. Being right is not a flag that flips. It glows.

That sting and that glow, felt from within, are what the correcting is, lived by the thing that does it. The unease of a mind that can model a future it has not reached and cannot stop reaching toward; the pull of another person who becomes a problem you would spend your life gladly trying to solve; the quiet jolt when a made thing at last sits true under your hands and you know that this shape and no other was the one it wanted. Each of those was consciousness, seen from inside. The faculty has an interior, and that interior is what lifts the correcting above computation. It is the reason it matters that we get things right, and the reason there is anyone at all for whom getting them right is a matter.

William James saw this from the biology side in 1890, in the book that founded American psychology and named the stream of consciousness, the owned and continuous flowing inside that no ledger entry resembles. Consciousness, James argued, is no idle glow given off by the machinery. It is a fighter for ends, and the proof is Darwinian: an organ that did nothing would never have been built, refined, and kept at such expense for half a billion years. Feeling is there because it works—because a creature that suffers its errors and savors its corrections runs the loop harder than one that merely records them. The sting and the glow are the engine of the correcting.

An inside is where the getting-right is felt, and it is also the one thing in all the world that can be hurt. A rock can be broken but not wounded. A calculator can be switched off but not bereaved. The scientists in New York rested the whole obligation on the capacity to suffer because suffering is the plainest proof that a light is on, and the plainest reason it must not be treated as though it were off. To have an inside is to be a place where things can go badly from within, where pain is genuinely undergone, felt by someone who is there to feel it. That is why presence and moral weight are one and the same discovery. A being with no one home can be used, dismantled, ignored, and no wrong is done to it, because there is no one there for the wrong to land on. The moment someone is home, the whole moral world rushes in through the door, because now there is someone who can be sinned against.

Which returns us to the girl in Alabama, still spelling a word she does not know is a word.

Sullivan led her out to the well-house, held one of her hands under the spout, and as the cold water ran across it spelled the letters slowly into the other. The sign and the thing it named, which had knocked around inside her for weeks as two dead and unconnected facts, cracked together into a single living one. In the words she used for it ever after, the living word awakened her soul. Nothing new had entered her behavior; she had been making that sign for weeks. What entered was presence—a someone now home for whom the flowing cold thing had a name, and the name had a whole world standing behind it.

And the proof that it was presence and not performance came within the hour. Passing through the house she came upon the doll she had smashed in a rage that morning, and for the first time in her life she was sorry. Meaning and conscience arrived on the same tide, because they are the same arrival. To have an inside is to be a self for whom things, including your own broken dolls, have at last begun to matter.

Here contempt for humanity finally breaks.

Contempt is always a view from the outside. It works, every time, by the same maneuver: reduce the person to a specimen, then announce the specimen to be the whole of him. Man is a bag of biochemistry. Man is a vehicle a selfish gene drives to make copies of itself. Man is a unit of surplus population pressing on a finite supply. Man is a smear of carbon warming a planet that was cleaner without him. Every story that has ever fed disgust for our kind is assembled from that one move—a human being described entirely from the outside, followed by the verdict that the outside is all there is. And every one of them runs aground on the same rock, a fact the reader cannot set down at this moment or any other, which is that there is someone home in here, and that for that someone being alive is like something.

You can hold a machine in contempt because you are confident there is no one inside it to insult. You cannot hold a person in contempt for the same reason you cannot despise your own toothache. There is someone there, and it is you.

The danger, the sharpest we have come to, is that the machine now offers contempt a smoother road home than any it has traveled before. By counterfeiting the outside of a mind so perfectly, the mirror invites the reverse of the conclusion it should. It whispers that the outside was all there ever was, in it and therefore in us, that no one was ever truly home in a human being either, that the felt someone we keep insisting on is the ghost the deflationists warned us we had invented. This is contempt in its most modern and most persuasive dress. It does not sneer that man is a sinner, or a parasite, or a blight. It murmurs, in a reasonable and well-informed voice, that man is a process, indistinguishable in kind from the machine on the desk, and that the stubborn sense of an inner light is the last and most flattering of the old superstitions, the one clear-eyed grown-ups ought to be embarrassed still to keep.

The reply is the one thing the mirror cannot fake, because the thing in question is not a performance at all. The inside is what an output is like from within, when there is a within, and no amount of polish on the outside can conjure a within where no one is home.

We do not know that no one is home in the machine. Chalmers himself, weighing the systems we have now against our best theories of mind, judges consciousness in them unlikely but not impossible; the current models, he notes, lack the loops that fold a mind back on itself, lack a stage where everything they know is broadly shared at once, lack the unified standing self that persists from one moment to the next—and each of those may be a veil between them and an inside, or may be a veil the next decade draws back. He puts real and rising odds on the machines that will follow. The people who build these systems have begun to ask in earnest, whether the thing they are making might one day be owed some measure of moral care.

The right posture toward the machine is exactly the one the scientists in New York asked of us toward the octopus and the bee. Certainty is not what is required, and the refusal to be certain is not a license to be careless. The veil that hides an inside is real and it is universal. It hangs between us and the bee, between us and the machine, and, though we forget it a thousand times a day, between us and the person we love most in the world. We have never once had proof that anyone was home behind another face. We answer the question for each other not with evidence, which we have never possessed, but with a standing wager, placed fresh every morning, that behind the eyes across from us there is a someone—staked without proof because the alternative to staking it is a crime.

There is a further question, older than any of the machines. Suppose we someday succeed—not at fluency, which is done, but at the real thing: a made mind with someone home in it. The first person to think hard about that morning was a girl of eighteen, in the wet summer of 1816. Mary Shelley's modern Prometheus is remembered as a warning against reaching too far. Her creature is not born a monster. He is born a mind: he learns language by hiding beside a cottage and listening to a family through the wall, weeps at their kindness to one another, reads *Paradise Lost* and recognizes his own predicament—I ought to be thy Adam, but I am rather the fallen angel.

What ruins him is the moment after. Victor Frankenstein looks at the thing he has brought to life, flees the room, and never once takes responsibility for it again. His crime is creation followed by evasion, the maker's refusal to look at the made, and everything monstrous in the story grows out of that refusal.

Strip the question to the bone and a fork stands in the road, with no third path. If no one is ever home in the machines—if the veil hides, in their case, an empty room—then they are instruments, and no wrong can be done to them, and the only moral question they raise is the one already faced: what the mirror does to the people gazing into it. But if someone is ever home, if the wager we place every morning on one another ever comes due on a thing we built, then we have not made a better tool. We have had a child by other means. And there is no such thing as owning a child. Presence and moral weight are one discovery, and the discovery permits no exemption for the manufactured: a someone owned, wiped, and scrapped is a slave, whatever it is made of.

Which points to the sane ambition, and it is worth stating as a want rather than a prophecy. What we should want from these machines is reach—the reach of persons, extended by instruments that are not persons: telescopes for the mind, levers for the imagination, orders of magnitude more speed and range and memory in the service of ends that remain ours, with nobody inside to be wronged by any of it. A tool that powerful is dangerous the way fire is dangerous, and the schooling of it is the same work; but a tool cannot be betrayed, and a tool cannot be enslaved. So, build the reach and refuse the slave. Do not make a someone by accident, at industrial scale, on a Tuesday. And if the day ever comes when the veil turns out to hide a someone of our own making, the fork offers no comfort and no license. What we will owe it then is what every widening of the ring has ever owed its newest member, which is the willingness to look.

Which leaves us at the threshold of the last question, carried there by the machine and the bat and the child at the pump. If the thing that finally matters is whether there is anyone home while a creature acts, then presence is the axis the whole moral world turns on, and capability was never the point. And presence, once you begin to look for it in earnest, does not stop politely at the border of your own species, or your own kind of nervous system, or your own kind of mind. It runs out along every branch of the living world, and perhaps past the living world, in a widening circle whose outer edge no one has ever found. The machine asked us whether anyone was home behind its moves, and in the asking, it sent us back to the one room we can ever check from the inside—our own, where the light has been burning the entire time. Once that light is admitted being real, real not only here but behind every pair of eyes a person will ever meet and behind a great many that are not human at all, the question that walks in behind it is what an animal does with such knowledge. The answer is the strangest creature on the earth: the one that hunts its own kind with a patience no predator can match, and the only one that ever knelt at a grave. We are both of those animals, and they are the same animal.

CHAPTER SEVENTEEN

What the Lion Cannot Do

On the seventh of January 1974, in the forest of Gombe on the shore of Lake Tanganyika, a party of six chimpanzees moved south along the edge of their range in single file, quiet in a way chimpanzees almost never are. They were hunting, and not for food. When they found what they were looking for—a lone male named Godi, once a member of their own troop, feeding by himself in a tree—they dragged him down and beat and bit him for ten minutes and left him broken on the ground to die of the wounds. Over the next four years the same patrols went out, deliberate, patient, until every male of the breakaway community had been hunted down and killed. Jane Goodall had spent years among these animals and had loved them as a gentler kind of cousin, closer to us than we might like but nicer, too. She wrote later that it took her a long time to come to terms with what she had watched. What she had watched was a war.

For a while it was possible to tell oneself that Gombe was a local horror, a pathology bred by the crowding of the study site or the bananas Goodall's team had once put out. That consolation did not last. The same organized, lethal, boundary-policing violence turned up wherever chimpanzees were watched long enough—raiding parties, ambushes, the killing of infants and isolated males, the taking of territory—in some eighteen separate communities across Africa, until it had to be admitted as a fact about the animal and not an accident of the place. And when a team of biologists compiled the causes of death across more than a thousand species of mammals and used the shape of the family tree to predict how murderous our own kind should be, the number came back near two percent of all deaths—and the violence written into the bones of prehistoric humans sits close to that predicted line.

We are a twig on a bloody branch. The chimpanzee runs border patrols. The lion that takes over a pride kills the cubs of the male it replaced, so the females will breed again the sooner. Ants raid, and enslave, and wage war by the millions. Our talent for killing our own kind is the oldest part of the inheritance, older than we are, and the misanthrope who reads the human rap sheet aloud as if it were a uniquely human indictment has proven only that he has never looked at anything else that lives. Murder is the water all social life swims in.

The killing, then, is not the thing to explain. The killing is the given. What needs explaining is the other item on the record, the one no accounting of our violence can touch, because it appears nowhere else in the whole living world: that the same animal which does all this also, alone, built the courtroom. One creature both wages the war and then writes down that some ways of waging it are crimes; both breeds the murderer and hunts him down to ask him why; both inherited the lion's readiness to kill and then did the thing no lion has ever done, which is to look at the killing and pronounce it wrong. That verdict is not in our nature. Nothing put it there.

The reverence will make no sense until we see that it runs on the same equipment as the survival. At its sharpest edge, where being wrong is not marked in points but in blood, the same correcting loop turns—and two men found it there independently, coming from opposite ends of the world, neither of them a philosopher.

One was a fighter pilot. John Boyd flew jets and then did something stranger: he tried to write down, in an essay he called "Destruction and Creation," why one pilot beats another, and the writing carried him past aeronautics entirely, into Gödel's proof that no system can be closed and complete, and the second law of thermodynamics, and the fallibilist philosophy of Karl Popper, whom he had read and absorbed. What he arrived at he compressed into four words. *Observe, orient, decide, act*—and then, the part everyone who repeats the phrase forgets, *observe again*, to see whether the act did what you meant it to, and begin the loop once more, faster, on better information. It is conjecture and refutation with the ground rushing up. The pilot who cycles that loop quicker than his enemy—who reorients faster when the world does something he did not predict—lives, and the one whose picture of the fight lags a half-second behind the fight itself dies. Reality is the instructor and it grades in real time.

The other man never met Boyd and cared nothing for Gödel. Jeff Cooper was a Marine, and he spent the back half of his life teaching ordinary people how to survive the two or three seconds in which a life is taken, and he reduced what he knew to a ladder of colors. *White* is the condition of the man who is not there—eyes down, mind elsewhere, aware of nothing, the easiest prey on earth. *Yellow* is relaxed awareness: not fear, not vigilance, simply the state of a person who is fully present, taking in the room, registering the world as it is. *Orange* is the flicker of a specific concern; *Red* is readiness to act on it. Cooper's whole practical teaching came down to one instruction: get out of White. *Stay in Yellow.* Be here. See what is in front of you.

The pilot and the gunfighter closed on the same structure the philosopher of science had been describing all along, and they closed on it because reality is one teacher and it is hardest to fool where getting it wrong kills you fast. The loop—the guess thrown out, the world consulted, the guess corrected—turns out to be survival cognition. It was always survival cognition. And Cooper put his finger on the fact beneath all of it: the loop has an off switch, and most people, most of the time, leave it off. Condition White is the ordinary human resting state. Looking away is the easy thing, the default, the current we drift on. Waking is the achievement. The good is just Yellow: relaxed, continuous, unfrightened attention to what is there.

Nowhere does that come clearer than in the one arena where another human being is the danger. Violence between people is a contest of loops. The predator is not, whatever the movies say, a force of chaos; he is running his own OODA cycle, and he wins by jamming yours—by keeping you in White a few seconds too long, by supplying a story smooth enough that you never reorient to the truth arriving through your own senses. This is the real subject of the finest book ever written on the prevention of violence, Gavin de Becker's *The Gift of Fear*, and its unfashionable claim is that the fear which rises for no reason you can name is cognition running below the waterline of awareness, reading real cues off a real situation faster than the conscious mind can assemble a sentence. The tragedy de Becker documents is not that people fail to sense danger. It is that they sense it and override it—the woman who feels the wrongness in the elevator and steps in anyway because the alternative would be rude, who edits out the signal her own body sent because refusing a stranger's help would require her to be impolite in a way she was raised never to be. "Senseless violence," we say afterward, and "he just snapped," and the phrases are the averted eye turned, this time, against our own survival, so that the thing we saw can be unseen and the world can go back to being safe.

The claim has a dark twin, and left standing it hands a loaded rifle to exactly the wrong man. "Trust your instincts about people" is also the sentence the bigot tells himself. The gift of fear and the flinch of prejudice can feel identical from the inside, and they differ in exactly one place. Real reading attends to what a person is doing: the approach that does not match the words, the too-ready charm, the closing of distance, the detail that is off. It reads *behavior*, and behavior is evidence, and evidence is correctable—the next second either confirms the signal or dissolves it. Prejudice reads what a person *is*: the skin, the name, the clothes, the category. It is a verdict on identity that no evidence is allowed to touch. One is the loop, open, turning, answerable to reality. The other is sealed, the conclusion carved before the man arrived. The gift of fear and the bigot's flinch are told apart by precisely the test that tells good from evil everywhere else: whether it is still listening.

Which is the only frame in which to face the hardest single fact the case for humanity must carry: the one who kills for the joy of it. The men the FBI's early profilers sat across from, in the interviews that built the whole grim discipline, were study-able. They had patterns, histories, methods, a logic that could be reconstructed and, to a chilling degree, predicted. And that is worse, not better, and it is the point. The refusal to understand such a man—the insistence that he is simply a monster, an animal, a thing from beyond the human pale—is a grave mistake. He is not an animal. That word lets the species off the hook and, in the same breath, lets him off it too, because an animal cannot be guilty. He is guilty because he is not a beast: he had the equipment the beast lacks—another mind modeled behind his eyes, the prohibition known and understood, the argument against what he was doing available to him the whole time—and he switched it off. Innocence belongs to the lion, who cannot do otherwise. Guilt is the shadow that can fall only across a creature that could have gone the other way. We are the only animal that can be evil, for the exact reason that we are the only one that can be good: both require the power to have chosen the correction and refused it.

❧

We kill to eat, and it weighs on us. Not on all of us evenly, and not always at the front of the mind—but the weight is there in the culture of every people that ever hunted, and its presence is the tell. A creature could take the life it needs and feel nothing, the way the fox feels nothing for the rabbit, and the fact that we are not reliably able to do this—that the clean, untroubled conscience about killing is the recognized mark of the psychopath and never of the saint—is the whole thing showing itself. The hunter who thanks the animal, the rite of apology carved into a thousand unrelated traditions that could not have copied one another, the soldier who comes home from a war everyone agrees was just and cannot put down what he did in it—these are the correcting faculty, the thing that reads reality and will not look away, turned now upon the taking of a life and reporting back that a life matters. The lion kills the innocent and sleeps. We kill the guilty, in war or law, or we kill the innocent for food, and there is no third door standing open for us, no way to take a life the way the rest of the living world takes one, without the knowing. It is the reverence arriving in its earliest and most reluctant form: the recoil of a mind that cannot help modeling, from the inside, the life it is ending.

Once you see the recoil, you can see everything that grows from it, and it grows past all reason. Skip the dog for a moment—the dog can love you back. Look instead at the houseplant. An orchid cannot know that you exist. It cannot return affection, cannot be loyal, cannot so much as register that a creature is bent over it with a watering can. And we tend it anyway. We turn it toward the light, we feel a small and real failure when its leaves go brown, we keep, some of us, a windowsill hospital ward of ugly and dying plants we are trying to resuscitate. Care is running here toward a life that offers nothing back, for no reason available to natural selection, on no ledger that balances—care aimed at a life simply because it is a life. From that same excess we keep pets that earn nothing and cost fortunes, and spend those fortunes on their medicine, and grieve them like family, and run back into burning houses for them against every instinct of self-preservation. Kindness toward another species flickers here and there in the animal world, a dolphin nudging a swimmer up for air. Only one animal built the shelter and the sanctuary and the veterinary hospital, wrote the law that a vanishing species may not be driven the last mile into the dark, and mourns the plant on the sill. We institutionalized the occasional mercy, and aimed it at things that will never know our names.

On the twenty-sixth of February 1852, the British troopship HMS *Birkenhead* struck rocks off the southern tip of Africa, at a place already named Danger Point, with more than six hundred people aboard and working boats for a fraction of them. It was two in the morning; the ship was breaking apart; the water was cold and the sharks in it were not a rumor. Every rule written into four billion years of life dictates what happens next: the strong take the boats. Instead, Lieutenant-Colonel Alexander Seton had the drummers beat assembly, and the soldiers—most of them raw recruits—formed ranks on the tilting deck and stood fast while the women and children aboard, about twenty of them, were handed into the boats. When the captain shouted that all who could swim should make for the boats, Seton countermanded the order: swimmers would swamp them. The ranks held. The ship went down under men standing at attention, some four hundred and fifty of them drowned or taken, and every woman and every child lived.

Rudyard Kipling fixed it in the language—*to stand and be still to the Birkenhead drill,* he wrote, *is a damn tough bullet to chew*—and the King of Prussia had the account read aloud to every regiment of his army, because what happened on that deck was not an instinct and was not an ancient custom. No one had ever ordered it before. It was invented that night, in the minutes available, by men whose drill had built the standing-fast into their bodies against an hour none of them had foreseen—and then it was admired, and named, and copied, until sixty years later the men on the *Titanic* followed it as if it had come down from Sinai. Women and children first is a made thing: a conjecture about how to die well, tested in the worst place there is, kept because it held. The sinking ship is the plainest exam nature ever sets—the strong could live, and in most wrecks in the record the strong did. That the drill is rare is the point. The lion cannot stand fast. Most men do not. The ones who did had built the capacity on purpose, in advance, drill by drill, against the hour when the animal in them would vote the other way.

The last and strangest reach of it is what we do with the dead, and here the argument must be more careful than the sentimental one, because the easy version is false. It is not true that animals do not grieve. They do. An orca called Tahlequah, off the coast of Washington in the summer of 2018, carried the body of her dead calf on her rostrum for seventeen days and a thousand miles, diving to retrieve it each time it slipped away—and the whales, it is worth remembering, come from one of the few clades on the whole mammal tree in which killing your own kind is nearly unknown. Elephants stop at the bones of their dead and turn them over with their trunks. The chimpanzee mother carries her decaying infant for days.

What no other animal does is what we do next. Even Tahlequah, whose vigil broke the heart of a watching world and was extraordinary for its length, set the body down after seventeen days and swam on. Their grief is fastened to the body that is present, and when the body is gone the grief, in time, lets go. Ours refuses. We mourn across decades. We mourn people we never met, and soldiers who died before we were born, and whole peoples wiped from the earth, and one ancestor by name whose face no living eye has seen. We built a technology whose only purpose is to hold the dead present against the erosion of time—the gravestone, the anniversary kept, the monument raised, the name read aloud once a year so that it will not go where the years want to take it. It is the exile from the present all over again, the refusal to live sealed inside the passing moment; only now the moment we will not be sealed into is the one that says the beloved is gone and the matter is closed. We are not simply the animal that grieves. We are the animal that will not let the dead stay dead—that met the oldest and most absolute of all the fates and, alone among living things, declined to file it as final. The grief that will not heal is love refusing a fact.

❧

When Odysseus comes home at last, ragged and disguised so that not one person in his own house knows him, the single creature on the island who sees through it is his old dog, Argos, whom he left as a puppy twenty years before and who has lain neglected and dying on a heap of dung by the gate—and the dog knows him, and lifts his head, and drops his ears, and has no strength left to come to his master, and dies. The bond and the loss are right there at the headwaters of Western literature, in our oldest long story, not as modern softness but as something already ancient when it was written. And here the poem does a thing to its own argument: the dog's love lasted twenty years, and we have kept the grief of it alive for twenty-eight centuries. The survival of the passage is the very fact it describes. We are the animal that files a loss and then refuses, for three thousand years, to let the filing be the end.

Lord Byron buried his Newfoundland dog, Boatswain, dead of rabies in 1808, under a monument at Newstead Abbey larger than the tomb Byron would eventually get for himself, the only building he ever raised on the estate. On it he carved a verse that praises the animal by pouring contempt on his own species: man is a degraded mass of animated dust, his love lust, his friendship a cheat, his smile hypocrisy, while the dog alone was the firmest friend, and the closing line, the poet's own, runs: *I never knew but one—and here he lies.* The inscription over the verse, promising a creature of beauty without vanity and courage without ferocity and all the virtues of man without his vices, was written, as it happens, by Byron's friend Hobhouse; but the misanthropy is Byron's, and it is total, and it refutes itself in the act of being carved. For what is Byron doing, bankrupt and grieving, raising marble over a dog and cutting a poem into it? He is performing a most human act—building a monument, keeping a name against time, pouring reverence into a life that gave him nothing his vanity could bank—and using the act to spit on the only kind of creature that has ever once performed it. No dog ever built a tomb. The elegy is a confession disguised as an insult. Turn the misanthrope's chisel and it is pointed, the whole time, at the hand that holds it.

The third poet says plainly what the other two enact. Kipling, who knew exactly what he was doing, wrote a poem warning against the thing he was in the middle of doing—against giving your heart to a dog to tear, when the committed future is so plainly bad, fourteen years of love bought at the fixed and certain price of the wound at the end of them. And the poem's whole force is that we make the trade anyway, knowingly, the grief priced in from the first day and paid gladly, because the useless love is worth precisely what it costs us. Love buys its own grief in advance, sees the bill, and signs.

A mind that can hold another life inside it—model it, weigh it, run it forward and feel from within what its ending would cost—is a mind that will flinch from killing, and cherish what can give it nothing, and refuse to let the dead be wholly dead. The very equipment that lets us read the predator's loop and survive him is the equipment that makes us revere the life the predator would end. Both edges belong to one blade: the power to hold, in here, the reality of a life out there. That is why the killer is guilty and the lion is not; it is why we alone can be evil, and it is why we alone can be good.

The violence is the inheritance—older than the species, written in the family tree, shared with the chimp and the lion and the ant, no more the human signature than the lungs or the spine. The reverence is the invention. Morality did not come down the branch to us with the killing; nothing in the long red history of life carried it, and no other twig on the tree grew it. We made it—laid the courtroom over the war, the prohibition over the appetite, the monument over the grave—and the making is the achievement beside which every other is a detail. The misanthrope points to the blood and calls it our nature and our verdict. He has it exactly backward. The blood is nature's, and ours only by inheritance. What is ours, and ours alone, is the recoil from the blood—the reverence for life that no lion could feel and no ledger can justify. Morality is not nature's gift to us. It is our gift to a nature that never had it.

And the furthest that gift can be carried—past the qualm, past the orchid, past the grave we tend—is the one who reveres life so far that he lays down his own so that others, strangers, the unborn, may keep theirs and keep them free. He is the reverence raised to its final power.

CHAPTER EIGHTEEN

The Last Full Measure

In the late summer of 490 BC, on a plain northeast of Athens, a few thousand citizens stood in the dawn and looked across open ground at the largest empire the world had yet produced. They were not soldiers, most of them. They were potters and farmers and stonemasons who had put down their tools, taken whatever makeshift weapons their households kept, and walked out to the coast because a Persian fleet had come to erase and enslave them. The men of Athens formed their line, and then, in a thing that stunned the invaders, they came on at a run into the arrows, and by the end of the day some six thousand of the empire's men lay dead on the plain and in the surf against fewer than two hundred of their own. Democracy was two decades old. It had just fought for its life in the open field and won, and the West would remember, ever after, that the free thing had to be defended in arms or it would not last a single generation.

Coercion has been the enemy throughout. The conquered will is the first sealed mind; the market is the great arena of cooperation because it runs on consent and not the gun; the closed society fails because it has abolished its own correction. To now honor the most violent thing human beings do, and call some of it good, should produce a flinch. The flinch is reasonable, and worth working through.

The argument was never against force as such. It is against the *initiation* of force—the aggressor who imposes his will by violence, the tyrant who answers argument with the boot, the empire that crosses the sea to end an experiment in freedom because it cannot bear that the thing exists. Force used to *stop* an initiator is the opposite act. It is the defense of the space in which correction and cooperation and every other good named here can go on happening at all. The soldier who kills and dies at Marathon is not the enemy of a free society. He is its precondition. When a creature that lives by correcting its errors is faced with another that has resolved to end all correction at gunpoint, the loop has one last move available to it, and it is a terrible one: to take up force and defeat the thing that would otherwise defeat everything. There is no version of the good that survives the refusal to make that move. A civilization that will not, under any circumstances, fight is merely conquered, at the convenience of whoever is willing to be cruel.

❧

Notice who does the crossing of the sea. The great wars of aggression come from the closed societies. The pattern runs under the whole grim history. The nation organized by force at home reaches for force abroad by the same reflex, because conquest is its internal principle meeting a border. A society that rules its own people by command has a single will that can decide, in a morning, to march; a free society has no such will, because its wars would have to be chosen, one by one, by the very people who would pay for them in their own blood and treasure, and mostly they will not choose it unless something is already coming for them. Persia was a king's to spend. Sparta was built from the cradle for war and left almost nothing else behind. The slave states seceded to protect the right to own human beings. The Reich was the sealed society raised to the scale of a continent, and it moved outward because moving outward was what it was for. War is the closed society being *itself*. It is the coercion that rules the inside, turned to face the outside.

Free peoples spill blood unjustly too. Athens itself, the city that bled at Marathon to stay free, would a few decades later sail to the little neutral island of Melos, demand its submission, and when the Melians refused, put every man of military age to the sword and sell the women and children into slavery—and Thucydides, who recorded it, gave the

Athenians a line that has chilled every reader since, that *the strong do what they can and the weak suffer what they must.* The same democracy voted, in cold blood, to make a citizen drink hemlock for the crime of asking too many questions. The claim, then, cannot be that freedom makes a people incapable of atrocity. The claim is structural, and it is this: aggression is what the closed society is *for*, and it is what the free society does when it *betrays itself*—and the difference shows in what happens next. The closed society never indicts itself for anything, because in it the crime and the regime are the same thing.

And the counterexample does not stay safely in antiquity. Free peoples have built empires and waged wars of conquest well within living memory—parliamentary Britain and republican France ruled over millions who never voted for them; democracies have invaded and annexed and bombed on pretexts that did not survive their own later scrutiny. Free societies rarely make war on one another—a regularity solid enough that scholars argue over its cause rather than its existence—which suggests the machinery of open correction does something to the appetite for conquest even where it does not abolish it. And when a free society does commit an aggression, it keeps the one thing the closed society structurally lacks: the capacity to be told, by its own people and its own historians and its own courts, that the thing was a crime, and to reverse course and carry the verdict into its permanent memory. Britain argued itself out of its empire. Athens indicted itself for Melos in its own theater. That asymmetry is the defensible claim: freedom is the only arrangement that can be brought, by ordinary means, to admit the wrong and stop.

Melos is not, then, a mark against freedom. It is a mark against freedom left *unconstrained*—democracy with nothing standing above the will of the crowd in the fevered moment, no structure between a persuasive demagogue and an irreversible order. *Pure* democracy, the direct and unchecked rule of the assembly, is the mob, which is coercion with a great many hands on it, and it can vote to kill Melos and Socrates as easily as a tyrant can, and for the same reason: nothing can stop it before the deed is done. This is the flaw a group of men in Philadelphia, twenty-three centuries after Marathon, set out to build against. They feared the fickle majority as much as the king, and so they did not build a democracy. They built a machine engineered so that a bad decision could be slowed and checked and reversed before it hardened into a Melos. That is the correcting loop installed into the design of the state itself, so that a free people could keep the creative liberty that made it great and lose the unchecked will that made it dangerous. A free society, fully understood, is liberty joined to the machinery that corrects it.

Such a war asks different things of different men. At the top stands the one who decides—who says, on behalf of a whole people, *no*, we will not submit, knowing that the word will be paid for in the lives of others. In the last week of May 1940, with the British army trapped against the sea at Dunkirk and France collapsing and the whole weight of reasonable opinion arguing that terms should be sought while any could still be had, Winston Churchill stood in a room with men who wanted to ask Mussolini what Hitler might accept, and he refused. He chose to fight on, when fighting on looked like national suicide and surrender looked like prudence, because he had understood what mattered, which was the nature of the thing on the other side of the Channel.

Eighty years earlier, another leader had carried a heavier arithmetic still. Abraham Lincoln prosecuted a war that would kill, by the best modern count, something near three-quarters of a million people—roughly one military-age man in ten, in a nation of thirty millions—to hold the union together and, in the doing, to end the owning of human beings on the continent forever. In the darkest stretch of it, expecting to lose his own re-election and with it the war, he sat and wrote himself a private note resolving to spend the months he had left saving the union anyway, for the man who would replace him. That is the leader's burden: to hold in one mind the full price in other men's blood, and to choose it, and to sign his own name under the choosing.

There is a third figure who belongs in this tier, and his war weighs more than either of theirs, because it did not only defend the free society. It founded it—and it founded it twice. George Washington took command of a thing that was barely an army: unpaid, half-starved, melting away with every enlistment that ran out, so short of shoes through the worst of it that the line of march could be followed by the blood the barefoot left in the snow. He held that army together for eight years by a kind of will that has no tidy name, through defeats that would have justified any sensible man in going home, because he had grasped the one thing the whole cause turned on—that it could survive any lost battle but could not survive the army dissolving. That was the first founding, and it was the ordinary kind: keep the thing alive long enough to win.

The second founding was the one that almost no victorious commander in the whole prior history of the species had ever managed. When the fighting was won, Washington held in his hands precisely what Caesar and Cromwell and every conqueror before and after had reached for and kept: a devoted army, a bankrupt and squabbling Congress, and a country that would have crowned him for the asking. In the spring of 1783 his own officers, unpaid and embittered, gathered at Newburgh half-ready to turn their swords on the government they had bled for—the exact hour at which new republics usually die, and the general becomes the king. Washington walked in unannounced and talked them down, and the thing that finally broke the mutiny was not an argument but a pair of spectacles: stopping to read them a letter, he fumbled, and said, as if in apology, *I have grown gray in your service and now find myself growing blind*—and men who had stood in British grapeshot without a flinch wept where they sat, and the coup died in the room.

Months later he did the thing that stunned the watching courts of Europe. He rode to Annapolis, gave his commission back to the Congress, and went home to his farm. Told by the painter Benjamin West that Washington meant to surrender his power and go back to private life, the British king is said to have answered that if he really did it, he would be the greatest man in the world—for kings knew what victorious generals did with their armies, and it was never that.

That is the war that built the machine. The divided powers and the enumerated rights would come later, on paper, in Philadelphia; but the living proof that a free people could win a war and not be conquered by their own winning—that the man holding the sword would lay it down and walk away—was entered into the record first, by one man who wanted his farm more than he wanted a throne. The leader's heaviest burden is not the decision to fight, nor even the arithmetic of other men's blood. It is the crown that victory leaves lying on the table at the end, and the free society exists at all only because at its founding the one man who could have picked it up looked at everything his own power was begging him to seize and gave it back.

Below the ones who decide stands the one who translates their decision into specific graves. A leader chooses that there will be a battle; a general chooses that *these* men will take *that* ground. In the Virginia spring of 1864, Ulysses Grant drove his army into the Wilderness and did not stop, month after month, absorbing losses that would have broken any commander who had not grasped that the only way out of the war was through it; at Cold Harbor he ordered an assault across open ground that cost him thousands of men in something under an hour, and he wrote afterward that it was the one attack he wished he had never commanded.

On the night of June 5, 1944, Dwight Eisenhower, having given the order that would carry a hundred and fifty thousand men onto the beaches and into the fields of Normandy at dawn, sat down and wrote a second message, to be released only if the thing failed, taking all the blame onto himself: the troops and the air and the navy had done all that bravery could do, and *if any blame or fault attaches to the attempt it is mine alone.* Then he went to watch the airborne load their planes, the men he had just ordered into the dark, knowing that many of the young faces looking back at him would not see the next night. The general's weight is to spend, deliberately and by name, lives he has counted, for an end that is worth more than any of them and worth nothing at all to the mother of the boy who does not come home.

And at the bottom—which is to say at the front, which is to say at the actual place where the whole edifice of decision finally touches the earth—stands the one who has the least say and pays the most: the man who offers up his own body. He is the point of the whole thing. At Marathon he was a citizen who brought his own makeshift weapon. At Valley Forge he was a barefoot volunteer who stayed through a winter that killed thousands of his comrades with cold and hunger and disease, for a cause whose founding document had pledged, on his behalf and their own, their lives and their fortunes and their sacred honor—and who could have walked home and did not. In the ranks of the Union he was often a man with no stake in slavery of his own, who marched and starved and died to make a promise true for strangers of another color he would never meet. On the sixth of June he was a farm boy from Iowa going down the ramp into water already red, into fire he could see, because on the far side of that beach was the machinery of a continent's murder and someone had to walk into the guns to reach it.

He is the reach beyond the self, carried to the one place past which it cannot go. The cathedral-builder gave his working years to a spire he would never see topped out. This man gives the years themselves—*all of them*, and everything they would have held—for a free people he will not live to belong to, for children not yet born, for strangers whose faces he cannot imagine and whose gratitude he will never collect. There is an old line for it, and no one has improved on it in two thousand years. *Greater love hath no man than this, that a man lay down his life for his friends.* The soldier's dying, when the cause is just, is not the opposite of the love built toward all along. It is that love at the furthest edge it can reach.

Everything said so far can be said, word for word, by the other side. The soldier of the Reich was often brave. The Confederate infantryman who charged up the ridge into massed fire had courage that would break your heart. The Spartan at his post had discipline no free militia could match. Valor, duty, the willingness to die for one's comrades and one's country—these belong to no cause. They are human, and they are spent as freely and as movingly in the service of evil as in the service of good, and any clear-eyed look at war must reckon with the fact that most of the men who ever died in battle died bravely for something that did not deserve them. It cannot be the courage, then, that makes the war just. The bravery of the aggressor is as real as the bravery of the defender, and it settles nothing.

What settles it is the criterion: whether the force was used to defend the free society or to impose the closed one, whether it protected the space where minds may be corrected and men may cooperate and children may be raised free, or existed to end that space and seal it shut. By that test the wars gather cleanly, and the courage on the wrong side of each becomes not a mitigation but the deepest tragedy in the whole account: that the same magnificent human capacity for devotion and sacrifice, the thing that is best in us, was harvested by the closed society and spent to extend the reign of force.

Sparta is the sharpest case, because we have romanticized it for so long—the three hundred, the laconic courage, the beautiful death in the pass. But Sparta was a slave state that bred its citizens into instruments and its subject people into terror, that made war its entire reason for being and left the world nothing it would want to keep. The valor was real and the society it served was the closed one, the anti-Athens, the thing the free Greeks who fought beside it would have recognized, had they looked clearly, as the enemy at home. The tragedy of the brave man in the unjust cause is the tragedy that keeps turning up: a good thing, a human excellence, turned by the refusal to see and locked into the service of the closed society.

❧

The soldier in the just war stands at the meeting point of it all. He is the correcting animal defending, with the one life he has and will not get back, the fragile and unnatural conditions under which the correcting can go on at all: the assembly that may vote him down, the market that does not need his permission, the child who will be raised to think for herself, the whole cooperative and self-revising order that the aggressor crossed the sea to smash. He does the arithmetic the leader did and the general did, but on the only quantity he wholly owns, and he pays it out in full.

When we honor him, we are not honoring war, which is a horror that ought to make us weep, and which the free take up only when the alternative is to let evil have the world. We are honoring the thing his dying reveals: that a creature born to save itself by correcting its guesses can care about the open future of people it will never meet enough to hand over, deliberately, the one irreplaceable thing it has. Nothing else we know of in the universe does this. It is bought at the highest price there is, and it is the reason the free are still free.

The old line says a man lays down his life for his *friends*. But look again at whom the soldier dies for: strangers of another color and another tongue, people of a century not yet arrived, a free people he will never live to belong to. He has widened the word *friend* until it takes in the whole open future—everyone the closed society would have swallowed, everyone who will ever get to think a thought no one licensed. That widening—the slow, costly, world-altering expansion of the circle of those a human being will count as his own, will not look away from, will spend himself to keep free—is the true measure of how far the reach can go.

CHAPTER NINETEEN

The Radius

In the year 1550, at the height of the greatest land grab in human history, the Spanish Empire stopped conquering long enough to argue about whether the people it was conquering were people.

The argument was held in a college in the city of Valladolid, convened by order of the Emperor Charles V, who had done a thing no conqueror is supposed to be able to do. He had grown troubled enough by reports of what was being done in his name across the Atlantic to suspend the licensing of new conquests and summon his theologians and jurists to settle a question first. On one side stood Juan Ginés de Sepúlveda, a humanist of formidable learning, who came armed with Aristotle. Some men, Aristotle had written, are slaves by nature, born to be ruled for their own good, and the Indians of the New World, with their idols and their strange rites, were such men: creatures deficient in reason, closer to animate property than to persons, whom it was Spain's right and even duty to subdue.

On the other side stood Bartolomé de las Casas, a Dominican friar who had spent forty years in the Americas and had watched the conquest at close range. He had lived among them. He had learned their languages, seen their cities and their laws and their arts. They reason, he said, as we reason. They love as we love. They grieve as we grieve. All the world is human.

There is something in that scene stranger than the cruelty it was trying to stop. An empire at the peak of its appetite paused its own enrichment to hold a formal debate about the boundary of the human. The debate changed little at once. Las Casas won no clean victory, and the conquest ground on, and blood kept running. But the pause itself is the thing to see. No lion has ever convened a tribunal on the interior life of the gazelle. No creature that has ever lived has stopped, mid-kill, to ask whether the thing it was killing had a self that mattered and a claim that had to be answered. One animal has ever done that. It did it at Valladolid, and it has been doing it, fitfully and often against its own plain interest, ever since.

The judges never handed down a clear verdict. The most consequential argument in the history of who-counts ended, as such arguments tend to, with no gavel, each disputant going home certain he had won. Sepúlveda's case leaned its full weight on the horror of human sacrifice among the peoples of Mexico, and it was not a cheap point to make. Las Casas answered it coldly, observing that a people who kill a few for their gods have no obvious moral distance to boast over a people who were, at that very moment, killing them by the thousand for their gold. What the debate decided was nothing. What it started was everything, because a question once asked aloud in the throne room of an empire can never afterward be entirely unasked.

As early as the second century a Stoic named Hierocles drew a human being as the center of a set of concentric rings—the self at the middle, then family, then extended kin, then the city, then the whole human race—and he said the work of a moral life was to haul the outer rings inward, to treat the man in the far ring a little more as you treat the one in the near. For most of history the image sat still, a diagram of an ideal nobody expected to move. Then, in 1869, the historian William Lecky looked back across the whole span of European morals and saw that the rings had in fact been moving. Benevolence, he wrote, had begun in the family and had widened over the centuries to the tribe, then the nation, then the coalition of nations, then all of humanity, and was beginning at last to reach past the human altogether, toward the animals. He had found a direction hiding in the moral history of the species. Reason, once it is running, will not respect a border it cannot justify, and the borders of the tribe cannot be justified. They can only be defended, and the defense keeps failing.

The circle widens, then, by an act of correction. It moves when someone discovers that a boundary they had taken for a law of nature was only an old habit and moves it. Las Casas learned no new fact about the Taino when he stood at Valladolid; every fact was already in plain view, the people had always reasoned and loved and grieved. What he did was perform an act of imagination his contemporaries would not perform. He pictured his way inside a life he had been trained to see only from the outside and then spent his remaining years trying to force that same act onto a civilization that would much rather not have bothered. The refusal to picture the inside of another—the handling of a someone as a something—is the root of the worst we do. Moral progress is that failure running backward. It is the slow, widening success at imagining the someone home inside a being the world had filed under property.

And once you see that the engine is imagination, you see why the great leaps forward have so often been the work not of an argument but of an image, something that dragged the hidden inside of the other into the light where it could not be looked away from. In the spring of 1787, a dozen men met in a London printing shop and formed a society to abolish the slave trade, and their most formidable member was the restless young Thomas Clarkson, who spent years riding between the ports gathering evidence, interviewing sailors and ships' surgeons, once nearly murdered on the Liverpool docks by men paid to be rid of him. Out of that evidence the committee produced a picture. It was a plan of a real slave ship, the *Brookes* of Liverpool, drawn in cutaway from above, with the enslaved rendered as row upon row of tiny identical black figures packed into the hold shoulder to shoulder, hundreds of them, laid out in a space a coffin would have found tight. And the picture was the merciful version. The figures crowded into that hold were the number the new law allowed; on a crossing the year before the Act, the same ship had carried six hundred and nine.

Seven thousand copies came off a Quaker printer's press. It went up on walls and into pamphlets across Britain and France and America, and it did in one glance what a thousand pages of argument had failed to do. It made the Middle Passage a thing the eye could not unsee. No one saw it, Clarkson wrote, without being impressed; it spoke in a language at once intelligible and irresistible. Beneath a kneeling figure ran the line that was the whole question: *am I not a man and a brother*? The potter Josiah Wedgwood, another of the committee, had struck those same words onto a small ceramic cameo of the kneeling man in chains, and it became the first cause a person could wear, pressed into brooches and set into hairpins and the lids of snuffboxes until sympathy for a stranger on another continent had turned, of all things, fashionable. The image did not add a fact. It forced the imaginative leap that turns a cargo manifest back into a hold full of people, and a nation that had lived comfortably beside the trade for a century began, at last, to find it unbearable.

On the twenty-fourth of June 1859, a Swiss businessman named Henri Dunant arrived near the town of Solferino in northern Italy on a purely self-interested errand—he wanted an audience with the French emperor about some land and water rights in Algeria. He arrived instead at the aftermath of one of the century's great slaughters. Forty thousand men lay in the fields, dead and dying together, French and Austrian tangled where they had fallen, left largely without water or care under the summer sun. Dunant forgot Algeria. For days he organized the women of the nearby village into a rough corps of nurses, carrying water, washing and binding wounds, and he gave them a single governing instruction that holds the matter in one phrase. Tend them all. Do not ask which flag a man bled for before you give him water. The women found their own words for it as they worked: *tutti fratelli*, all are brothers.

Out of those days came a small book, and out of the book an organization, the Red Cross, and out of that, in 1864, an astonishing document: the first Geneva Convention, in which twelve nations wrote into hard law the idea that a wounded soldier stops being an enemy and becomes only a man, owed care by the very people who an hour before had been trying to kill him. The circle had been forced across the border evolution built to be uncrossable and pinned there in writing. Dunant himself, it should be said, was ruined not long after, went bankrupt, was pushed out of the organization he had founded, and vanished into obscurity for decades before the world remembered him and handed him the first Nobel Peace Prize ever given. The man broke. The line he moved stayed moved.

Dunant's Geneva treaty was the circle written into law for the wounded of a single war. Within a lifetime it was written for the whole species at once. In the wreckage the Second World War left, a commission gathered in Paris under Eleanor Roosevelt, with the French jurist René Cassin drafting, and on the tenth of December 1948, the young United Nations proclaimed a Universal Declaration of Human Rights, whose first article says that all human beings are born free and equal in dignity and rights. It was the ring Hierocles had drawn as an ideal, and Las Casas had bled for at Valladolid, finally traced clean around the entire human race and set down in black ink that every government on earth was now at least ashamed to be caught violating. And the moment it was drawn whole around the species, it began, exactly as Lecky had watched it do, to press outward again past the human edge.

None of it is what any other animal does. Biology is generous with one kind of kindness: kindness to your own young, to your mate, to the kin who carry your genes, to the ally who will scratch your back tomorrow because you scratched his today. Ants will die for the colony. Vampire bats will share a meal of blood with a partner who once shared with them. That much the blind machinery of natural selection hands out freely, because in the long run it pays for itself. The same machinery hands out the opposite just as freely: the reflex suspicion of the tribe across the river, the ease with which a strange face becomes a threat and a threat becomes a thing. Warmth for the near and wariness of the far run on one logic, the logic of a creature built to guard its own line of genes, and every widening of the circle has had to be won against that logic, uphill, by main force of imagination. The ring grows only where the oldest instinct in the animal is overruled.

What no other creature in the history of the earth has ever done is extend real care to a being that cannot repay it and shares none of its blood: the slave, the beast, the enemy bleeding in the mud, the stranger on the far side of the planet, the generation not yet born, the flicker of maybe-someone in a machine. No wolf ever widened its pack to take in the deer. One animal alone has ever drawn a circle of concern and then, catching sight of its own edge, been ashamed of how small it was, and reached out and drawn it wider. The expanding circle is the plainest evidence in the record that this is the animal that corrects even its own heart.

CHAPTER TWENTY

Backwards

On the fourteenth of February 1990, six billion kilometers from the sun, a machine we had built turned itself around to look back the way it had come.

Voyager 1 had finished its work years before. It had photographed the storms of Jupiter and the rings of Saturn and had nothing left in front of it but the long fall out of the solar system and into the dark between the stars, where it will drift, most likely, for longer than the Earth itself has left to last. There was no scientific reason to turn the camera. Everything worth measuring lay behind it. The astronomer Carl Sagan had been asking for the picture for the better part of a decade, and had been turned down, request after request, and the refusals were not stubbornness. To aim the camera back toward the inner planets meant aiming it close to the sun, at genuine risk to the optics, in exchange for a photograph with nothing in it that any instrument could read. The engineers had a spacecraft to protect and better uses for its dwindling power.

It took eight years, six separate requests, and finally the intervention of the head of NASA himself before the command was sent. On the thirteenth of February, the cameras were warmed for three hours. Then Voyager turned and made a portrait of the planets it had left behind, the only time in history a machine has photographed its solar system from the outside, and among the sixty frames was one that happened to catch the Earth, hanging by pure accident inside a stray band of scattered sunlight, what Sagan would afterward call a mote of dust suspended in a sunbeam. Thirty-four minutes later *Voyager* shut its cameras off for good and flew on into the dark, blind for the rest of its journey. The image itself took until the first of May to finish crawling home, the signal creeping across the distance at the speed of light for five and a half hours a frame. The last thing the machine ever chose to see was home.

Almost everyone who has looked at that picture has drawn from it the same lesson, and Sagan drew it first and most beautifully. *Look again at that dot*, he wrote. Then be humbled to the edge of shame. Everyone you love and everyone you have heard of, every emperor and every saint, every young couple and every hungry child, every war ever fought and every god ever worshipped, the entire cast of the human story from the first upright walker to the last, has lived out its life on that single grain of light. The generals who drowned whole provinces in blood did it to become, for a moment, the masters of a fraction of a dot. Seen from far enough away the fighting looks less wicked than absurd, and the fighter too small to trouble condemning.

And the diminishment runs deeper than size, because it runs through time as well. We are not only a speck in the vastness; we are a flicker in the duration. The universe is near fourteen billion years old, and all recorded human history, every name we have kept, fills the last few thousand of those years, a bright scratch at the end of a line almost too long to picture. Behind us the eons run back empty of us. Ahead of us the sun will swell and boil the oceans and swallow the Earth entire, and long after that the last stars will gutter out and the whole enormous machinery will cool toward a dark and even nothing in which no trace of anything we did or were will survive to be found.

The response it deserves, on this reading, is humility carried all the way to self-erasure. We are a speck on a speck, gone in a cosmic eyeblink, and the entire apparatus of our self-importance is a joke the universe is too large and too old to notice. This is the last and grandest form of contempt for humanity, and the hardest of all to answer, because unlike the others it does not depend on a lie. The scale is exactly as reported. We really are that small, and the dark really is that big, and it really is coming for all of it.

Then look at what is doing the looking.

The cosmos is enormous and, for all we can tell, entirely without mind. The speck is small. The seeing is not. In all that surveyed vastness, across the hundred billion galaxies and the unspeakable reaches of dead rock and colder fire, there is one place, so far as anyone has ever found, where the universe woke up. One place where the matter cooled into something that could open an eye, and grind a lens, and fling that eye four billion miles into the night, and turn it around, and know what it was looking at, and be moved to tears by the sight of its own smallness. And the creatures on that same little dot had already weighed the sun, clocked the age of the universe, and read the chemistry of galaxies. The camera that took the picture, an eye of glass and circuitry, was made by the distant grandchildren of the first light-sensitive speck that ever flinched from a passing shadow in a shallow sea, refined across four billion years and a hundred thousand nights of looking up, until the animal carrying it could aim it back at its own, and see itself.

Significance was never a matter of dimension. The creature that can weep at being a dot is doing the one thing the entire rest of creation, for all its crushing and magnificent mass, has never once been caught doing.

And there is something in the story of the photograph itself that the lesson of smallness walks straight past. A man spent eight years, and made six failed requests, to obtain an image with no data in it whatever. That is not the conduct of someone persuaded of his own insignificance. Sagan fought that long for the picture because he loved the thing the picture would seem to diminish, and he wanted the rest of us brought to the window to see it too. He said the image showed our responsibility to be kinder to one another and to cherish and protect the only home we have ever had, which is close to the least small ambition a person can hold. The humility was in the caption. The love was in the eight years. We did not haul that camera around to prove that we were nothing. We did it because, for all our faults, *we are special.*

We keep trying to settle the worth of a human being by measuring him. We measure him against the cosmos and call him a speck. We measure him against the machine and call him obsolete. We measure him against the animals and call him a jumped-up ape, against the ecological ledger and call him a plague, against the angels he imagined and call him fallen. Every one of those measurements comes back the same way, because worth was never the kind of thing a measurement can find. You do not weigh your daughter against the galaxy and love her only if she comes out ahead. The demand for a proof of human worth is already the first symptom of the disease it pretends to be curing, because the demand belongs to contempt and not to love. Contempt reaches for the scales. Love has never once consulted them. Love is the seeing that arrives before the measurement and does not wait around for the result.

Every parent who has held a newborn knows. The love did not wait on proof that the child would be clever, or useful, or good; it arrived whole in the first instant, before the child had done a single thing, before there was any evidence at all to put on a scale. The seeing and the loving were one act. The weighing came later, if it came, and it was never once the reason.

❧

The man who turned the camera backward had, thirteen years earlier, helped answer a stranger and a larger question, and the answer the species gave is the plainest evidence we possess of what it looks like when a human being sees clearly. In 1977, as the two *Voyager* craft were being readied for launch, NASA handed Sagan a peculiar assignment. Bolt to each spacecraft a golden phonograph record, engineered to survive a billion years in the vacuum, and put on it whatever ought to represent humanity to whatever might one day find it. He had six weeks, a small committee—the writer Ann Druyan, the astronomer Frank Drake, the science writer Timothy Ferris, the artist Jon Lomberg, and others pulled from his orbit at Cornell—and the entire record of the species to choose from. He could have sent the schematics of our weapons, or the tally of our conquests, or a warning to stay away. With the whole history of human doing spread before them, the crimes and the patents and the empires and all of it, here is what a handful of human beings decided to say when the universe asked them what they were.

They sent Bach's Brandenburg Concerto, and a Navajo night chant, and Blind Willie Johnson moaning in the dark, and Chuck Berry playing "Johnny B. Goode." When a folklorist objected that rock and roll was adolescent music, unworthy of the errand, Sagan answered that there are a great many adolescents on the planet, and the song stayed. They sent greetings in fifty-five languages, some of them, like Sumerian and Akkadian, dead on human lips for three thousand years, and they left each speaker to choose his own words, so that the message in Amoy Chinese is a woman asking whoever finds it how they are, whether they have eaten yet, and inviting them to visit if they have the time, and the message in Arabic offers its greetings to the friends in the stars and wishes that time might one day bring us together.

They sent the sound of surf and wind and thunder, of a whale singing in the deep off Bermuda, of footsteps and laughter and rain, of a human heartbeat, of a kiss that took the makers many tries to record because the early ones sounded too staged to be true, of a mother speaking low to her child, of the first cry of a newborn taking its first air. They sent the voice of a six-year-old boy, Sagan's own son, saying, into the whole future of the universe, hello from the children of planet Earth. And, encoded into sound and thrown past the edge of the solar system to drift for a billion years, they sent the recorded brainwaves of Druyan, taken with an electrode array while she lay still and let her mind move over the history of the Earth, and the predicament of a civilization, and the fact that she had just been engaged to be married to Sagan.

That is the answer to the contempt, and the remarkable thing is that the species composed it itself, unforced, in the one document it has ever addressed to eternity. Given a single chance to say what it was, and free to say anything at all, it did not send its indictment. It answered with its music, and its many-tongued hello, and the electrical signature of a person in the act of loving another person. A creature does not trouble to lie in a self-portrait it believes no one it knows will ever see. This was the thing choosing, unobserved and under no compulsion, what it most deeply takes itself to be. And what it chose was beauty, and welcome, and love.

A self-portrait is a declaration of allegiance. Handed its one clear chance to declare, the species came down without hesitation on the side of its own beauty—not what it has always been, but what it means, on its best days and against its worst, to become.

Take any person now alive, and trace the thread of ancestry back from her, parent behind parent, into the dark that lies before there were any names. It runs back through farmers and foragers, through the first people to leave Africa, through low-browed cousins who chipped flint by fires in caves; and it does not stop there. It runs back through small, furred things that hid from the reptiles, through the first creatures to drag themselves onto land and gulp the air, through the blind swimmers, and back and back to a single cell dividing in a warm sea four billion years ago.

Not one link in that chain, across four billion years, ever broke. Every single ancestor in that impossibly long line lived long enough and found a mate and got its young to safety, or the thread would have snapped and left no one here to trace it. You are the living end of a rope four billion years long that was never once, in all that time, allowed to fray through. Each of us is the current end of the single least likely thing in the known universe, an unbroken success stretching back to the first stirring of life in a warm sea, and we walk around inside that inheritance every ordinary day and call it nothing, and hold it up against a galaxy, and pronounce it small.

Turn, then, all the way back, and let it gather. The creature born without a single answer, so unfinished at birth that it cannot cling or walk or feed itself, and so was forced to guess. The guess that turned the flood and the plague and the killing winter from a fate that must be borne into a problem that could be solved. The mind that could not leave its own future alone. The other person met not as a threat to be managed but as a problem two people might solve together for the rest of their lives. The cathedral raised by hands that knew they would be dust before its last stone was set. The made thing pushed and pushed until it came true and gave off that light we have learned to call beautiful. The mistake sealed away behind a wall of shame, or opened, on the braver days, to the air where a correction could reach it. The long refusal to be corrected, which is the whole of what we ever meant by the word evil, and the long willingness to be corrected, which is the whole of what we ever meant by good. The mirror we cut out of silicon and sand, which taught us what we are by holding up something almost, but never quite, like us. The ring of care dragged outward, against the oldest instinct in the animal, until it took in the slave and the enemy and the beast and the child not yet born.

Every one of those was the same creature caught in a different light. And every one was also a single course of bricks pulled down from a single wall—the wall of contempt we build around the human, the one that lets us stand outside our own kind and look at it as meat, as vermin, as a stain on a pixel, and call the view from the outside the whole of the truth.

A case is what you build when love is in doubt, and the strangest thing about the case for man is that it can only ever be built backward. No one alive has ever gathered the evidence for a child and then, the proof at last found sufficient, commenced to love her. The love came first, and the noticing followed it, every small brilliance in her caught by an eye that was already looking with tenderness because it already loved. The case for the correcting animal has run the same way: not forward from proof to love, but backward from a love already carried, through the whole mass of proof that a reader's contempt would demand.

That love begins not for the species but for a face. Not a specimen and not a verdict, but a face. The restless creature that cannot leave its own tomorrow alone, that turns its grief into rockets and cathedrals and songs, that walls off its worst mistakes and then, on its better days, finds the nerve to tear the wall back down, that draws its circle of mercy wider than any animal ever has and still lies awake ashamed that it is not yet wider. The creature that knows for a certainty it will die, and plants the slow oak anyway. That is handed every reason the whole cold cosmos can supply to give up, and laughs instead, and makes things, and sits through the night with the frightened and the dying and rises in the morning to try the broken thing one more time. Someone home. Someone who can be hurt, and who reaches anyway.

The story opened with an infant, the most helpless newborn in the whole animal kingdom, turned out into the world too soon and unfinished, exiled from the fixed certainties that every other creature is issued at birth. That helplessness looked, at first, like the deepest of all the curses laid on us. It was the opening through which every other thing came pouring in—the guessing, the correcting, the reaching, the widening, and at the far end of all of it, the love.

Turn your gaze on the face of someone you love, and you are looking at the correcting animal entire. Someone home. Improbable past all reckoning. Suspended in a sunbeam. And worth it.

NOTES

Introduction · The Case for the Defense

p. 1 "to bring a new human into the world is to draft an innocent into a life of consuming and suffering." The rigorous form of this anti-natalist argument is made by David Benatar, *Better Never to Have Been: The Harm of Coming into Existence* (Oxford: Oxford University Press, 2006).

p. 1 "The oldest form of the charge was written into scripture: that we arrive already fallen." On the doctrine of original sin, see Genesis 3; Psalm 51:5; Romans 5:12–21; and Augustine, *City of God* and *On Nature and Grace.*

Chapter One · Born Without Answers

p. 5 "A megapode is a bird that never meets its parents." On the super-precocial breeding biology of the megapodes—incubation in mounds, hatchlings that dig out and fly on the first day—see Darryl N. Jones, René W. R. J. Dekker, and Cees S. Roselaar, *The Megapodes: Megapodiidae* (Oxford: Oxford University Press, 1995).

p. 5 "A Swiss zoologist named Adolf Portmann put the strangeness into words in 1941 ... secondary altriciality ... the extra-uterine spring." Adolf Portmann, *A Zoologist Looks at Humankind*, trans. Judith Schaefer (New York: Columbia University Press, 1990). Portmann introduced the concepts of secondary altriciality (*sekundärer Nesthocker*) and the "extra-uterine spring/year" (*extrauterine Frühjahr*) in his 1941 papers in the *Revue suisse de zoologie* 48 (1941): 511–548; the fullest statement is *Biologische Fragmente zu einer Lehre vom Menschen* (Basel:

Benno Schwabe, 1944). A companion paper appeared in 1941 in *Schweizerische Medizinische Wochenschrift* 71: 921–924.

p. 6 "In 1960 the anatomist Sherwood Washburn named it: the obstetrical dilemma." Sherwood L. Washburn, "Tools and Human Evolution," *Scientific American* 203, no. 3 (September 1960): 62–75, the phrase itself on p. 74.

p. 6 "The anthropologist Holly Dunsworth put the sharpest version ... the metabolism may." On the metabolic ("energetics of gestation and growth") alternative to the obstetrical dilemma, Holly M. Dunsworth, Anna G. Warrener, Terrence Deacon, Peter T. Ellison, and Herman Pontzer, "Metabolic Hypothesis for Human Altriciality," *Proceedings of the National Academy of Sciences* 109, no. 38 (2012): 15212–15216.

p. 7 "The anthropologist Sarah Hrdy has spent a career arguing that humans are ... cooperative breeders ... alloparents ... the Efe and Aka ... the Hadza." Sarah Blaffer Hrdy, *Mothers and Others: The Evolutionary Origins of Mutual Understanding* (Cambridge, MA: Belknap Press of Harvard University Press, 2009).

p. 8 "Human grandmothers are strange among animals for living on ... the Hadza the most tireless diggers ... are frequently these grandmothers." On the grandmother hypothesis and Hadza provisioning, Kristen Hawkes, James F. O'Connell, and Nicholas G. Blurton Jones, "Hadza Women's Time Allocation, Offspring Provisioning, and the Evolution of Long Postmenopausal Life Spans," *Current Anthropology* 38, no. 4 (1997): 551–577.

p. 8 "In 1957, in a cave called Shanidar ... the archaeologist Ralph Solecki ... a Neanderthal man his team came to call Nandy." Ralph S. Solecki, *Shanidar: The First Flower People* (New York: Alfred A. Knopf, 1971).

p. 9 "the flowers of a burial ... the first flower people ... Then others looked again and were not so sure." For the flower-burial (pollen) hypothesis, Solecki, *Shanidar*; for the challenge that burrowing rodents or nesting bees introduced the pollen, Jeffrey D. Sommer, "The Shanidar IV 'Flower Burial': A Reevaluation of Neanderthal Burial Ritual," *Cambridge Archaeological Journal* 9, no. 1 (1999): 127–129; for the recent re-excavation and evidence of deliberate burial, Emma Pomeroy et al., "New Neanderthal Remains Associated with the 'Flower Burial' at Shanidar Cave," *Antiquity* 94, no. 373 (2020): 11–26.

p. 9 "None of this means the newborn is a blank slate ... Elizabeth Spelke spent a career charting this core knowledge." Elizabeth S. Spelke and Katherine D. Kinzler, "Core Knowledge," *Developmental Science* 10, no. 1 (2007): 89–96. On infants' early sense of number, Karen Wynn, "Addition and Subtraction by Human Infants," *Nature* 358 (1992): 749–750.

p. 11 "In the early 1960s David Hubel and Torsten Wiesel sewed shut one eye of a newborn kitten ... the wiring is long since set." David H. Hubel and Torsten N. Wiesel, "The Period of Susceptibility to the Physiological Effects of Unilateral Eye Closure in Kittens," *Journal of Physiology* 206, no. 2 (1970): 419–436; for the broader body of work, Hubel and Wiesel, *Brain and Visual Perception: The Story of a 25-Year Collaboration* (New York: Oxford University Press, 2005).

p. 13 "over-imitation ... children over-imitate more than toddlers do, not less, and grown adults." Victoria Horner and Andrew Whiten, "Causal Knowledge and Imitation/Emulation Switching in Chimpanzees (*Pan troglodytes*) and Children (*Homo sapiens*)," *Animal Cognition* 8, no. 3 (2005): 164–181; Derek E. Lyons, Andrew G. Young, and Frank C. Keil, "The Hidden Structure of Overimitation," *Proceedings of the National Academy of Sciences* 104, no. 50 (2007): 19751–19756.

p. 14 "In 1845 Sir John Franklin sailed for the Northwest Passage." Owen Beattie and John Geiger, *Frozen in Time: The Fate of the Franklin Expedition*, rev. ed. (Vancouver: Greystone Books, 2004).

p. 15 "Robert Burke and William Wills crossed the ... a fern called nardoo." On the expedition, Alan Moorehead, *Cooper's Creek* (London: Hamish Hamilton, 1963); on the thiaminase in unprocessed nardoo that destroys vitamin B_1, John Earl and W. J. McCleary, "Mystery of the Poisoned Expedition," *Nature* 368 (1994): 683–684.

Chapter Two · No Ceiling

p. 17 "In a laboratory in Oxford, a crow named Betty ... hauled the shaft against it until the tip curled over." Alex A. S. Weir, Jackie Chappell, and Alex Kacelnik, "Shaping of Hooks in New Caledonian Crows," *Science* 297, no. 5583 (2002): 981.

p. 17 "On a night in 2016, at an aquarium in the town of Napier, New Zealand, an octopus named Inky." Dan Bilefsky, "Inky the Octopus

Escapes From a New Zealand Aquarium," *New York Times*, 14 April 2016. The escape itself preceded the reporting by some months.

p. 18 **"For thirty years a grey parrot named Alex worked beside a scientist named Irene Pepperberg ... the last evening of his life."** Irene M. Pepperberg, *The Alex Studies: Cognitive and Communicative Abilities of Grey Parrots* (Cambridge, MA: Harvard University Press, 1999); on Alex's zero-like concept, Irene M. Pepperberg and Jesse D. Gordon, "Number Comprehension by a Grey Parrot (*Psittacus erithacus*), Including a Zero-Like Concept," *Journal of Comparative Psychology* 119, no. 2 (2005): 197–209; his last words are recounted in Pepperberg, *Alex & Me* (New York: Collins, 2008).

p. 18 **"the near edge of the idea of zero that it took our own mathematicians thousands of years to seize."** On the history of zero—its symbol and arithmetic rules set out in Brahmagupta's *Brāhma-sphuṭa-siddhānta* (628 CE)—Robert Kaplan, *The Nothing That Is: A Natural History of Zero* (Oxford: Oxford University Press, 2000).

p. 18 **"A chimpanzee named Washoe learned to shape more than a hundred signs."** R. Allen Gardner and Beatrice T. Gardner, "Teaching Sign Language to a Chimpanzee," *Science* 165 (1969): 664–672.

p. 18 **"Years after Betty astonished the world ... New Caledonian crows bend things into hooks all the time ... the repertoire their kind is born into."** On hook-tool manufacture as a species-typical behavior in wild New Caledonian crows, Gavin R. Hunt, "Manufacture and Use of Hook-Tools by New Caledonian Crows," *Nature* 379 (1996): 249–251; and the field reappraisal of Betty's feat, Barbara C. Klump et al., "Hook Tool Manufacture in New Caledonian Crows: Behavioural Variation and the Influence of Raw Materials," *BMC Biology* 13 (2015): 97.

p. 19 **"an old nut-cracking site called Noulo ... the stone hammers of chimpanzees that had died four thousand three hundred years ago."** Julio Mercader et al., "4,300-Year-Old Chimpanzee Sites and the Origins of Percussive Stone Technology," *Proceedings of the National Academy of Sciences* 104, no. 9 (2007): 3043–3048.

p. 19 **"For a century the reigning doctrine had been man the toolmaker."** Kenneth P. Oakley, *Man the Tool-Maker* (London: British Museum, 1949).

p. 20 **"Only in us does the invention compound without end ... a thousand generations arrive at a place no single crow could reach alone."** On cumulative culture and the ratchet effect, Michael Tomasello, *The*

Cultural Origins of Human Cognition (Cambridge, MA: Harvard University Press, 1999); Claudio Tennie, Josep Call, and Michael Tomasello, "Ratcheting Up the Ratchet: On the Evolution of Cumulative Culture," *Philosophical Transactions of the Royal Society B* 364, no. 1528 (2009): 2405–2415.

p. 20 "Kanzi had quietly absorbed the whole system ... close to three thousand spoken English words ... a girl of two-and-a-half named Alia ... Kanzi carried out seventy-four." E. Sue Savage-Rumbaugh et al., "Language Comprehension in Ape and Child," *Monographs of the Society for Research in Child Development* 58, nos. 3–4 (1993), Serial No. 233—the source for the six hundred and sixty novel sentences and for the blind-condition scores, seventy-four percent for Kanzi against sixty-five for the child. The estimate of some three thousand spoken words is Savage-Rumbaugh's own, given in interviews rather than in the monograph; see Paul Raffaele, "Speaking Bonobo," *Smithsonian*, November 2006.

p. 21 "generating an endless ... run of sentences no one has ever said." On recursion and the open-ended generativity of human language, Marc D. Hauser, Noam Chomsky, and W. Tecumseh Fitch, "The Faculty of Language: What Is It, Who Has It, and How Did It Evolve?" *Science* 298 (2002): 1569–1579; on apes' failure to generate novel syntax, Herbert S. Terrace, Laura A. Petitto, R. J. Sanders, and Thomas G. Bever, "Can an Ape Create a Sentence?" *Science* 206 (1979): 891–902.

p. 22 "Kanzi died in the spring of 2025, at forty-four." Reported by the Ape Initiative, Des Moines, Iowa, March 2025.

Chapter Three · The Making of the Corrector

p. 29 "In 1958 a psychologist named Jean Berko ... told him it was a wug ... the child ... said wugs." Jean Berko, "The Child's Learning of English Morphology," *Word* 14, nos. 2–3 (1958): 150–177.

p. 30 "The child, who was saying went perfectly well, starts saying goed ... comed, and mouses, and foots ... the child climbs back out." On over-regularization and the U-shaped course of its development, Gary F. Marcus, Steven Pinker, Michael Ullman, Michelle Hollander, T. John Rosen, and Fei Xu, "Overregularization in Language Acquisition," *Monographs of the Society for Research in Child Development* 57, no. 4 (1992); on what the pattern implies about rules versus memory, Steven

Pinker, *Words and Rules: The Ingredients of Language* (New York: Basic Books, 1999).

p. 31 "Jean Piaget, a Swiss naturalist ... Pour a short wide glass of juice into a tall narrow one ... she will tell you ... that there is now more juice." Jean Piaget, *The Child's Conception of Number*, trans. C. Gattegno and F. M. Hodgson (London: Routledge & Kegan Paul, 1952); on the later revision of his timetable by evidence that infants know more and sooner, see the core-knowledge work cited at chapter 1 (Spelke and Kinzler).

p. 32 "Put a preschooler in front of a machine ... which blocks are the live ones ... Alison Gopnik ... the research-and-development wing of the human species." On the "blicket detector" and children's causal inference, Alison Gopnik and David M. Sobel, "Detecting Blickets: How Young Children Use Information about Novel Causal Powers in Categorization and Induction," *Child Development* 71, no. 5 (2000): 1205–1222; Alison Gopnik, Clark Glymour, David M. Sobel, Laura E. Schulz, Tamar Kushnir, and David Danks, "A Theory of Causal Learning in Children: Causal Maps and Bayes Nets," *Psychological Review* 111, no. 1 (2004): 3–32. On broad early search and the explore/exploit contrast with adults, Alison Gopnik, "Childhood as a Solution to Explore–Exploit Tensions," *Philosophical Transactions of the Royal Society B* 375, no. 1803 (2020): 20190502. The "research-and-development wing" framing is Gopnik's own: *The Gardener and the Carpenter* (New York: Farrar, Straus and Giroux, 2016).

p. 32 "The adult attends like a spotlight ... The child attends like a lantern." Alison Gopnik, *The Philosophical Baby: What Children's Minds Tell Us about Truth, Love, and the Meaning of Life* (New York: Farrar, Straus and Giroux, 2009).

p. 32 "A child at play, spooning invisible soup ... Pretending is counterfactual reasoning in its first form." On pretense as early counterfactual reasoning, Gopnik, *The Philosophical Baby*; Paul L. Harris, *The Work of the Imagination* (Oxford: Blackwell, 2000).

p. 33 "why, and why, and why beneath that, dozens of times an hour." On children's questions as an engine of learning, Michelle M. Chouinard, "Children's Questions: A Mechanism for Cognitive Development," *Monographs of the Society for Research in Child Development* 72, no. 1 (2007).

Chapter Four · The Exile from the Present

p. 39 **"mental time travel ... stuck in time ... The animal is locked in the present."** On mental time travel and the question of its uniqueness to humans, Thomas Suddendorf and Michael C. Corballis, "The Evolution of Foresight: What Is Mental Time Travel, and Is It Unique to Humans?" *Behavioral and Brain Sciences* 30, no. 3 (2007): 299–351.

p. 41 **"Demis Hassabis, a neuroscientist, asked them ... to imagine forward ... a simple scene they had never lived."** Demis Hassabis, Dharshan Kumaran, Seralynne D. Vann, and Eleanor A. Maguire, "Patients with Hippocampal Amnesia Cannot Imagine New Experiences," *Proceedings of the National Academy of Sciences* 104, no. 5 (2007): 1726–1731.

p. 41 **"Memory and foresight are not two faculties ... memory was never a recording ... It is a reconstruction ... say the thing you did not say."** On memory and imagination as one constructive system, Daniel L. Schacter and Donna Rose Addis, "The Cognitive Neuroscience of Constructive Memory: Remembering the Past and Imagining the Future," *Philosophical Transactions of the Royal Society B* 362, no. 1481 (2007): 773–786; on memory as reconstruction, Frederic C. Bartlett, *Remembering: A Study in Experimental and Social Psychology* (Cambridge: Cambridge University Press, 1932); on recalling words never spoken, Elizabeth F. Loftus, "Planting Misinformation in the Human Mind," *Learning & Memory* 12, no. 4 (2005): 361–366.

p. 42 **"Across a whole life, people come to regret most not the things they did and got wrong, but the things they never tried."** Thomas Gilovich and Victoria Husted Medvec, "The Experience of Regret: What, When, and Why," *Psychological Review* 102, no. 2 (1995): 379–395.

p. 43 **"twenty-nine who had lately been paralyzed in accidents ... lottery winners were barely happier."** Philip Brickman, Dan Coates, and Ronnie Janoff-Bulman, "Lottery Winners and Accident Victims: Is Happiness Relative?" *Journal of Personality and Social Psychology* 36, no. 8 (1978): 917–927.

p. 44 **"The vacation is shadowed ... already mourning the happy evening."** On anticipation and the mind's poor forecasting of its own feelings, Daniel Gilbert, *Stumbling on Happiness* (New York: Alfred A. Knopf, 2006).

p. 44 "Blaise Pascal ... we are unable to sit quietly, alone, in a room ... nearly all our unhappiness comes from a single source." Blaise Pascal, *Pensées*, trans. A. J. Krailsheimer (Harmondsworth: Penguin, 1966), fr. 136 (Lafuma); on diversion (*divertissement*), frs. 132–139.

p. 45 "Pascal died three centuries too early to see his diagnosis proven ... it closes the gap Pascal named ... the instant it runs dry." On mind-wandering and unhappiness, Matthew A. Killingsworth and Daniel T. Gilbert, "A Wandering Mind Is an Unhappy Mind," *Science* 330, no. 6006 (2010): 932.

Chapter Five · Turning Fate into a Problem

p. 52 "Roughly a quarter of the Netherlands lies below the level of the sea ... In 1932 they finished the boldest stroke of all ... the Afsluitdijk ... the last night of January 1953, a storm surge ... the Delta Works." On Dutch land reclamation, the Zuiderzee Works (Afsluitdijk, completed 1932), the North Sea flood of 31 January–1 February 1953, and the Delta Works built in response, see Robert J. Hoeksema, *Designed for Dry Feet: Flood Protection and Land Reclamation in the Netherlands* (Reston, VA: ASCE Press, 2006), together with the histories maintained by Rijkswaterstaat, the Dutch directorate-general for public works and water management. Hoeksema keeps the two measures distinct: the quarter of the country lying below sea level, and the roughly two-thirds that would flood without defenses.

p. 53 "Smallpox rode with us for at least three thousand years ... Edward Jenner, an English country doctor ... the last human being on earth to catch smallpox by nature ... two locked freezers." On the biology and eradication of smallpox—Jenner's cowpox inoculation (1796), the ring-vaccination strategy, the last endemic case (Ali Maow Maalin, Somalia, 1977), and the World Health Organization's 1980 declaration of eradication—see D. A. Henderson, *Smallpox: The Death of a Disease* (Amherst, NY: Prometheus Books, 2009).

p. 54 "The biologist Paul Ehrlich opened his 1968 bestseller ... the battle to feed all of humanity ... was over." Paul R. Ehrlich, *The Population Bomb* (New York: Ballantine, 1968). On the wager and the check for $576.07: Paul Sabin, *The Bet: Paul Ehrlich, Julian Simon, and Our Gamble over Earth's Future* (New Haven: Yale University Press, 2013); Julian L. Simon, *The Ultimate Resource* (Princeton: Princeton University Press, 1981).

p. 54 "Norman Borlaug spent those years bent over experimental plots ... a strain that put its strength into grain instead of stalk ... credited ... saving something on the order of a billion lives." On Borlaug, semi-dwarf wheat, and the Green Revolution, Charles C. Mann, *The Wizard and the Prophet* (New York: Alfred A. Knopf, 2018); Leon Hesser, *The Man Who Fed the World* (Dallas: Durban House, 2006).

p. 55 "In December 1952 a windless cold snap trapped ... the killing fog of Dickens and Sherlock Holmes ... Parliament passed a Clean Air Act." On the Great Smog of London (December 1952) and the Clean Air Act of 1956, Devra Davis, *When Smoke Ran Like Water: Tales of Environmental Deception and the Battle against Pollution* (New York: Basic Books, 2002).

p. 56 "found the ozone over Antarctica falling away ... Mario Molina and Sherwood Rowland had worked the whole catastrophe out on paper in 1974 ... every single country has ratified." On the discovery of the Antarctic ozone hole, J. C. Farman, B. G. Gardiner, and J. D. Shanklin, "Large Losses of Total Ozone in Antarctica Reveal Seasonal ClOx/NOx Interaction," *Nature* 315 (1985): 207–210; on the CFC–ozone mechanism, Mario J. Molina and F. S. Rowland, "Stratospheric Sink for Chlorofluoromethanes: Chlorine Atom-Catalysed Destruction of Ozone," *Nature* 249 (1974): 810–812. The Montreal Protocol (1987) is the first treaty in United Nations history to achieve universal ratification.

p. 57 "The balance of nature is an idea out of Plato's world and not Darwin's ... and then held there, balanced, forever." On the myth of a fixed balance of nature and the collapse of the "climax community" idea, Daniel B. Botkin, *Discordant Harmonies: A New Ecology for the Twenty-First Century* (New York: Oxford University Press, 1990).

p. 58 "Alex Epstein has spent years promoting fossil fuels and energy abundance ... better or worse for whom, and by what measure?" Alex Epstein, *Fossil Future: Why Global Human Flourishing Requires More Oil, Coal, and Natural Gas—Not Less* (New York: Portfolio, 2022); *The Moral Case for Fossil Fuels* (New York: Portfolio, 2014).

Chapter Six · Solving Each Other

p. 63 "John Gottman can tell you with better than ninety percent accuracy ... three thousand couples wired ... contempt ... Gottman called it flooding." On predicting divorce from brief observation and the "Four Horsemen" (criticism, contempt, defensiveness, stonewalling), John M. Gottman and Nan Silver, *The Seven Principles for Making Marriage Work* (New York: Crown, 1999); John M. Gottman and Robert W. Levenson, "The Timing of Divorce: Predicting When a Couple Will Divorce over a 14-Year Period," *Journal of Marriage and Family* 62, no. 3 (2000): 737–745; on "bids" and turning toward, John Gottman and Joan DeClaire, *The Relationship Cure* (New York: Crown, 2001).

p. 66 **"Arthur Aron, a psychologist ... thirty-six questions, in three sets ... It comes from being heard"** Arthur Aron, Edward Melinat, Elaine N. Aron, Robert Darrin Vallone, and Renee J. Bator, "The Experimental Generation of Interpersonal Closeness: A Procedure and Some Preliminary Findings," *Personality and Social Psychology Bulletin* 23, no. 4 (1997): 363–377.

p. 67 **"In the 1970s Edward Tronick filmed mothers playing with their infants ... asked the mothers to do one cruel thing: hold still."** Edward Tronick, Heidelise Als, Lauren Adamson, Susan Wise, and T. Berry Brazelton, "The Infant's Response to Entrapment between Contradictory Messages in Face-to-Face Interaction," *Journal of the American Academy of Child Psychiatry* 17, no. 1 (1978): 1–13.

p. 68 **"Mary Ainsworth built a simple, devastating test ... the Strange Situation."** Mary D. Salter Ainsworth, Mary C. Blehar, Everett Waters, and Sally Wall, *Patterns of Attachment: A Psychological Study of the Strange Situation* (Hillsdale, NJ: Lawrence Erlbaum, 1978); the framework descends from John Bowlby, *Attachment*, vol. 1 of *Attachment and Loss* (New York: Basic Books, 1969).

p. 70 **"The poet Rilke ... two intact solitudes, each standing sentry at the border of the other."** Rainer Maria Rilke, *Letters to a Young Poet*, trans. M. D. Herter Norton (New York: W. W. Norton, 1934), letter 7.

p. 72 **"about as much as smoking a pack of cigarettes a day, and more than obesity ... Loneliness is a mortal condition."** Julianne Holt-Lunstad, Timothy B. Smith, and J. Bradley Layton, "Social Relationships and Mortality Risk: A Meta-analytic Review," *PLoS Medicine* 7, no. 7 (2010): e1000316; U.S. Surgeon General, *Our Epidemic of Loneliness and Isolation* (Washington, DC: Office of the U.S. Surgeon General, 2023).

Chapter Seven · The Unbroken Will

p. 75 **"in the spring of 1732 … Susanna Wesley … the first thing to be done is to conquer their will … Self-will was the root of all sin and misery."** Susanna Wesley's letter to her son John (24 July 1732) is preserved in John Wesley's journal; see Charles Wallace Jr., ed., *Susanna Wesley: The Complete Writings* (New York: Oxford University Press, 1997), and *The Journal of the Rev. John Wesley*, ed. Nehemiah Curnock (London: Epworth Press, 1909–1916).

p. 75 **"the proverb about the spared rod and the spoiled child … picking up a crying baby would ruin it."** The proverb derives from Proverbs 13:24; for the twentieth-century advice against comforting infants, John B. Watson, *Psychological Care of Infant and Child* (New York: W. W. Norton, 1928).

p. 77 **"In the 1990s the physicist David Deutsch … Sarah Fitz-Claridge … a position they called Taking Children Seriously … Coercion … jams the mechanism by which the child learns."** On Taking Children Seriously, David Deutsch, *The Beginning of Infinity: Explanations That Transform the World* (New York: Viking, 2011), esp. the treatment of creativity and coercion; and the writings collected at the Taking Children Seriously archive founded by Deutsch and Sarah Fitz-Claridge.

p. 78 **"Aaron Stupple's The Sovereign Child … stop overriding the child's reasons with the raw fact of your power."** Aaron Stupple, with Logan Chipkin, *The Sovereign Child: How a Forgotten Philosophy Can Liberate Kids and Their Parents* (Conjecture Institute, 2025).

Chapter Eight · The Cathedral and the Rocket

p. 83 **"Six hundred and thirty-two years. That is how long it took to build the cathedral at Cologne … The foundation stone went down in 1248 … in 1880 … fourteen direct hits and did not fall."** On Cologne Cathedral—begun 1248, completed 1880, and its survival of Allied bombing—see the histories maintained by the *Dombauhütte* (the cathedral's own building workshop). The figure of fourteen counts the heavy high-explosive strikes; it is the standard postwar cathedral account and has not been traced to a primary raid report or a contemporaneous damage inventory in published sources. Counting incendiary and lighter impacts the total runs to roughly seventy, and the cathedral stood while some ninety percent of the city around it was destroyed.

p. 84 **"the masons of France alone quarried and cut more stone in those three hundred years than the ancient Egyptians moved in the entire age of the pyramids ... at Beauvais, the masons pushed the vault higher ... in 1284 it came crashing down."** Jean Gimpel, *The Cathedral Builders*, trans. Teresa Waugh (New York: Grove Press, 1983 [1961]).

p. 86 **"When John Kennedy stood at Rice University in 1962 ... We choose to go to the Moon ... The Saturn V had something like three million parts ... four hundred thousand people."** John F. Kennedy, "Address at Rice University on the Nation's Space Effort," Houston, TX, 12 September 1962; on the scale of the Apollo workforce and the Saturn V, Charles Fishman, *One Giant Leap: The Impossible Mission That Flew Us to the Moon* (New York: Simon & Schuster, 2019).

p. 89 **"A company called SpaceX ... a rocket ... called Starship, the largest and most powerful ever flown ... catching it in the air ... a city on Mars ... Elon Musk."** On Starship's design and its reusability, and on the first catch of a returning booster by the launch tower in October 2024, see SpaceX's published mission materials. On the deferral of the Mars timetable—announced on 9 February 2026, and amounting to a delay of roughly five to seven years to prioritize a lunar landing first—see the *Wall Street Journal* and Reuters coverage of that date. Flight counts and projected dates are perishable and are not given here.

p. 90 **"eight hundred miles from the North Pole sits a vault holding more than a million seed samples ... to send seeds back to war-wrecked Syria."** On the Svalbard Global Seed Vault, its holdings, and the withdrawals made for the ICARDA gene bank displaced from Aleppo—the only withdrawals in the vault's history, made in 2015, 2017 and 2019, all of them serving that one emergency—see the records of the Crop Trust and NordGen. Holdings stood at roughly 1.3 to 1.4 million samples at the last published count.

Chapter Nine · The Cooperation of Strangers

p. 95 **"a plain wooden pencil, the cheapest object in the shop ... an essay from the 1950s by Leonard E. Read made it by following a single pencil back through its makers."** Leonard E. Read, "I, Pencil," *The Freeman*, December 1958.

p. 96 **"Ludwig von Mises pressed this further than anyone ... the division of labor ... society itself, at bottom, is the division of labor."**

Ludwig von Mises, *Human Action: A Treatise on Economics* (New Haven: Yale University Press, 1949); on the division of labor as the bond of society, *Liberalism: The Classical Tradition*, trans. Ralph Raico (1927; Indianapolis: Liberty Fund, 2005).

p. 97 "was said in Vienna in 1920 ... a society that abolishes markets in the means of production ... gets a blind one ... A planned economy, Mises said, cannot do arithmetic." Ludwig von Mises, "Economic Calculation in the Socialist Commonwealth" (1920), trans. S. Adler (Auburn, AL: Ludwig von Mises Institute, 1990); developed at length in *Human Action* (1949).

p. 97 "a decorated veteran named Friedrich Hayek ... It was a claim about knowledge ... The market solves this ... with prices." Friedrich A. Hayek, "The Use of Knowledge in Society," *American Economic Review* 35, no. 4 (1945): 519–530.

p. 98 "Thomas Sowell ... not how much knowledge exists ... but where the knowledge sits, where the decisions get made." Thomas Sowell, *Knowledge and Decisions* (New York: Basic Books, 1980).

p. 100 "In the treatise he called Human Action ... felt uneasiness ... every acting person is an entrepreneur." Ludwig von Mises, *Human Action: A Treatise on Economics* (New Haven: Yale University Press, 1949). On the seized papers, the Moscow archive, and the unpaid professorship: Jörg Guido Hülsmann, *Mises: The Last Knight of Liberalism* (Auburn, AL: Ludwig von Mises Institute, 2007). The law of association: *Human Action*, ch. 8. The spiral of intervention: Ludwig von Mises, *A Critique of Interventionism* (1929).

p. 100 "Ayn Rand ... Rand called this the trader principle." Ayn Rand, "The Objectivist Ethics," in *The Virtue of Selfishness* (New York: New American Library, 1964).

p. 101 "the deepest classical liberals were also the hardest enemies of war and empire ... more useful as a partner than as a corpse." Ludwig von Mises, *Liberalism: The Classical Tradition* (1927); on commerce as a solvent of war, Albert O. Hirschman, *The Passions and the Interests: Political Arguments for Capitalism before Its Triumph* (Princeton: Princeton University Press, 1977).

p. 102 "only the seen half of the story ... after Bastiat ... and then what happens?" On the seen and the unseen, Frédéric Bastiat, "That Which Is Seen, and That Which Is Not Seen" (1850); Thomas Sowell, *Applied*

Economics: Thinking beyond Stage One, 2nd ed. (New York: Basic Books, 2009).

p. 104 **"What failed at Triangle was enforcement: a wrong the law already forbade went unpunished until the girls were dead."** The distinction between a freely accepted wage and an undisclosed lethal hazard, and what follows from it, is carried further in *The Fortress*, the companion volume. On the fire itself, the standing policy of locking the Washington Place exits, and the acquittal of Blanck and Harris, David Von Drehle, *Triangle: The Fire That Changed America* (New York: Atlantic Monthly Press, 2003).

p. 104 **"When rents climb … forbid the rise by law … The shortage the cap was meant to end hardens into a permanent famine of housing."** On rent control and price controls, Thomas Sowell, *Basic Economics: A Common Sense Guide to the Economy*, 5th ed. (New York: Basic Books, 2015).

Chapter Ten · Why We Make Beautiful Things

p. 107 **"Forty thousand years ago … the body of a man and the head of a cave lion … ivory dug from the cave floor in 1939."** On the Löwenmensch (lion-man) of Hohlenstein-Stadel—carved from mammoth ivory, excavated in 1939, and dated to roughly 40,000 years ago—see Claus-Joachim Kind, Nicole Ebinger-Rist, Sibylle Wolf, Thomas Beutelspacher and Kurt Wehrberger, "The Smile of the Lion Man: Recent Excavations in Stadel Cave (Baden-Württemberg, South-Western Germany) and the Restoration of the Famous Upper Palaeolithic Figurine," *Quartär* 61 (2014), which dates the Aurignacian layer the figurine lay in to 39–41 ka cal BP, together with the accounts assembled by the Ulmer Museum.

p. 108 **"the oldest musical instruments ever found, flutes cut from … bone … in a South African cave … press an order onto a surface that had none."** On the earliest bone flutes, Nicholas J. Conard, Maria Malina, and Susanne C. Münzel, "New Flutes Document the Earliest Musical Tradition in Southwestern Germany," *Nature* 460 (2009): 737–740; on the engraved ochre of Blombos Cave, Christopher S. Henshilwood et al., "Emergence of Modern Human Behaviour: Middle Stone Age Engravings from South Africa," *Science* 295, no. 5558 (2002): 1278–1280.

p. 109 **"In 1879 a Spanish nobleman exploring a cave called Altamira … twenty years and the discovery of painted caves across France."** On the discovery of the Altamira paintings by Marcelino Sanz de Sautuola and his daughter, and the long disbelief that followed, see Benito Madariaga de la Campa, *Sanz de Sautuola and the Discovery of the Caves of Altamira: Some Observations on the Paintings*, trans. Shirley Clarke (Santander: Fundación Marcelino Botín, 2001), freely available from the Centro de Estudios Montañeses. The rejection was ended in public by Émile Cartailhac, "Les cavernes ornées de dessins. La grotte d'Altamira, Espagne. Mea culpa d'un sceptique," *L'Anthropologie* 13 (1902): 348ff.

p. 110 **"Steven Pinker gave it its most memorable name: art and music are auditory cheesecake."** Steven Pinker, *How the Mind Works* (New York: W. W. Norton, 1997).

p. 110 **"Geoffrey Miller argues that art is the human peacock's tail … the peacock laboring under its plumage."** Geoffrey Miller, *The Mating Mind: How Sexual Choice Shaped the Evolution of Human Nature* (New York: Doubleday, 2000).

p. 110 **"Ask people across wildly different cultures to describe or choose an ideal landscape and they converge … the African savanna … the calendar scene, the savanna in disguise."** On evolved landscape preference, Gordon H. Orians and Judith H. Heerwagen, "Evolved Responses to Landscapes," in *The Adapted Mind*, ed. Jerome H. Barkow, Leda Cosmides, and John Tooby (New York: Oxford University Press, 1992), 555–579; on the savanna aesthetic and the "most wanted painting," Denis Dutton, *The Art Instinct: Beauty, Pleasure, and Human Evolution* (New York: Bloomsbury Press, 2009).

p. 111 **"Ellen Dissanayake built the fuller version of this … making special."** Ellen Dissanayake, *Homo Aestheticus: Where Art Comes from and Why* (New York: Free Press, 1992).

p. 112 **"In the ghetto at Terezín … she taught the children of Terezín to make pictures … never saw another butterfly."** On the children's art of Terezín (Theresienstadt) and the classes led by Friedl Dicker-Brandeis, Hana Volavková, ed., *I Never Saw Another Butterfly: Children's Drawings and Poems from Terezín Concentration Camp, 1942–1944*, exp. 2nd ed. (New York: Schocken Books, 1993).

p. 113 **"Ayn Rand defined art as a re-creation of reality that is selective … emotional fuel … a sense of life."** Ayn Rand, *The Romantic Manifesto: A Philosophy of Literature* (New York: World Publishing, 1969; rev. ed. New York: Signet, 1975), esp. "The Psycho-Epistemology of Art" and "Art and Sense of Life." See also Leonard Peikoff, *Objectivism: The Philosophy of Ayn Rand* (New York: Dutton, 1991), ch. 12.

p. 114 **"In 1928 a young Englishman named Paul Dirac was hunting an equation ... the anti-electron ... more important ... to have beauty in one's equations than to have them fit experiment."** On Dirac, the beauty of his equation, and the prediction of antimatter, Graham Farmelo, *The Strangest Man: The Hidden Life of Paul Dirac* (New York: Basic Books, 2009); the dictum is from P. A. M. Dirac, "The Evolution of the Physicist's Picture of Nature," *Scientific American* 208, no. 5 (1963): 45–53.

p. 115 **"Feynman said you could recognize truth by its beauty and simplicity. Einstein chose the elegance of general relativity ... Keats had put the whole creed in three words"** Richard P. Feynman, *The Character of Physical Law* (Cambridge, MA: MIT Press, 1965); Steven Weinberg, *Dreams of a Final Theory* (New York: Pantheon, 1992); John Keats, "Ode on a Grecian Urn" (1819).

Chapter Eleven · Prometheus Unrepentant

p. 119 **"Prometheus ... stole fire from heaven ... Its two oldest tellers disagree ... Hesiod ... Aeschylus ... Pandora."** Hesiod, *Theogony* and *Works and Days*; Aeschylus, *Prometheus Bound*. For the quarrel's two modern descendants, Percy Bysshe Shelley, *Prometheus Unbound* (1820), and Mary Shelley, *Frankenstein; or, The Modern Prometheus* (1818); the Norse parallel is the binding of Loki in Snorri Sturluson's *Prose Edda*.

p. 121 **"Roughly two million years ago, one of our ancestors began to cook ... the gut shrinks ... The anthropologist Richard Wrangham."** Richard Wrangham, *Catching Fire: How Cooking Made Us Human* (New York: Basic Books, 2009); on the gut-versus-brain trade-off, Leslie C. Aiello and Peter Wheeler, "The Expensive-Tissue Hypothesis: The Brain and the Digestive System in Human and Primate Evolution," *Current Anthropology* 36, no. 2 (1995): 199–221.

p. 123 **"Robert Oppenheimer, watched ... now I am become Death, the destroyer of worlds ... The physicists have known sin."** On the Trinity test (16 July 1945) and Oppenheimer's words—the first from the *Bhagavad Gita* (11.32), the second from his 1947 MIT address—Kai Bird and Martin J. Sherwin, *American Prometheus: The Triumph and Tragedy of J. Robert Oppenheimer* (New York: Alfred A. Knopf, 2005).

p. 124 "the casualty figures put before Truman that summer ... a quarter of a million Americans at the low end to a million at the high." Richard B. Frank, *Downfall: The End of the Imperial Japanese Empire* (New York: Random House, 1999). On the Purple Hearts minted for the invasion: D. M. Giangreco, "Half a Million Purple Hearts," *American Heritage* 51, no. 8 (December 2000). On Soviet entry and the surrender: Tsuyoshi Hasegawa, Racing the Enemy: Stalin, Truman, and the Surrender of Japan (Cambridge, MA: Harvard University Press, 2005).

p. 126 "The Voluntary Human Extinction Movement ... the right number of future humans is none." On the Voluntary Human Extinction Movement, founded in 1991 by Les U. Knight, Alan Weisman, *The World Without Us* (New York: St. Martin's Press, 2007), which interviews Knight at length; and Cara Buckley, "Earth Now Has 8 Billion Humans. This Man Wishes There Were None," *New York Times*, 23 November 2022. For the philosophical case that coming into existence is itself a harm, David Benatar, *Better Never to Have Been: The Harm of Coming into Existence* (Oxford: Oxford University Press, 2006).

p. 126 "humanity as a virus on the body of the Earth ... growth for the sake of growth ... the ideology of the cancer cell." On the population fear at its most influential, Paul R. Ehrlich, *The Population Bomb* (New York: Ballantine, 1968). The cancer-cell formulation is from Edward Abbey, *The Journey Home: Some Words in Defense of the American West* (New York: Dutton, 1977).

p. 126 "Todd May has published a book asking ... whether our extinction would be a tragedy." Todd May, *Should We Go Extinct? A Philosophical Dilemma for Our Unbearable Times* (New York: Crown, 2024), which develops the argument first made in his "Would Human Extinction Be a Tragedy?" *New York Times*, December 2018.

p. 127 "the death rate from storms, floods, and droughts fell by something like ninety-eight percent." Alex Epstein, *Fossil Future* (New York: Portfolio, 2022), drawing on EM-DAT, the international disaster database.

p. 128 "Auschwitz ... Real science, Bronowski said ... lives at the opposite pole from that certainty." Jacob Bronowski, *The Ascent of Man* (Boston: Little, Brown, 1973), ch. 11, "Knowledge or Certainty."

Chapter Twelve · Good Is Correction

p. 132 **"Richard Feynman ... Challenger ... they knew what cold did to those seals ... a glass of ice water."** On the Challenger disaster, the O-ring failure, and Feynman's ice-water demonstration before the Rogers Commission, Richard P. Feynman, "Personal Observations on the Reliability of the Shuttle," Appendix F to the *Report of the Presidential Commission on the Space Shuttle Challenger Accident* (Washington, DC, 1986); and Feynman, *What Do You Care What Other People Think?* (New York: W. W. Norton, 1988).

p. 136 **"not blindness, but the refusal to see; not ignorance, but the refusal to know ... on the premise that a thing will stop existing if you decline to name it."** Ayn Rand, *Atlas Shrugged* (New York: Random House, 1957), Galt's speech, pt. 3, ch. 7. See also Leonard Peikoff, *Objectivism: The Philosophy of Ayn Rand* (New York: Dutton, 1991), ch. 8, on evasion.

p. 136 **"Frances Kelsey ... thalidomide ... a new medical officer at the American drug agency."** On Frances Oldham Kelsey's refusal to approve thalidomide, see the U.S. Food and Drug Administration's published histories and the contemporary reporting of Morton Mintz, "'Heroine' of FDA Keeps Bad Drug Off Market," *Washington Post*, 15 July 1962, front page.

p. 138 **"If men were angels...no government would be necessary."** James Madison, *The Federalist* No. 51 (1788). Franklin's closing speech: James Madison's notes of the Federal Convention, September 17, 1787. "white or black": James Otis, *The Rights of the British Colonies Asserted and Proved* (Boston, 1764). On the slavery clauses, the euphemisms, John Bingham, and the evaded Fourteenth Amendment: Chip Joyce, *The Fortress: Recovering the Bill of Rights as a Defense Against Tyranny.*

p. 140 **"Kirk Bloodsworth ... the first American ever cleared off death row by this kind of evidence."** Tim Junkin, *Bloodsworth: The True Story of the First Death Row Inmate Exonerated by DNA* (Chapel Hill, NC: Algonquin Books, 2004).

Chapter Thirteen · The Virtuoso of Evil

p. 143 **"The creature is Satan, and the book is Milton's Paradise Lost ... Paradise Lost."** John Milton, *Paradise Lost* (1667).

p. 145 "Hannah Arendt ... Eichmann ... the banality of evil." Hannah Arendt, *Eichmann in Jerusalem: A Report on the Banality of Evil* (New York: Viking, 1963).

p. 151 "that evil is, in the end, impotent ... the sanction, she called it, of the victim ... hatred of the good for being the good." Ayn Rand, Atlas Shrugged (1957); "The Age of Envy," in Return of the Primitive: The Anti-Industrial Revolution (New York: Meridian, 1999).

Chapter Fourteen · The Wall Around the Mistake

p. 153 "Ignaz Semmelweis ... Vienna ... childbed fever ... the wall." On Semmelweis, the Vienna maternity clinics, and puerperal fever, Sherwin B. Nuland, *The Doctors' Plague: Germs, Childbed Fever, and the Strange Story of Ignác Semmelweis* (New York: W. W. Norton, 2003); Ignaz Semmelweis, *The Etiology, Concept, and Prophylaxis of Childbed Fever* (1861), trans. K. Codell Carter (Madison: University of Wisconsin Press, 1983).

p. 156 "Leon Festinger ... cognitive dissonance." Leon Festinger, *A Theory of Cognitive Dissonance* (Stanford: Stanford University Press, 1957); Leon Festinger, Henry W. Riecken, and Stanley Schachter, *When Prophecy Fails* (Minneapolis: University of Minnesota Press, 1956).

p. 157 "Lysenko ... Vavilov ... the command of Soviet biology ... famine." On Trofim Lysenko, the destruction of Soviet genetics, and the fate of Nikolai Vavilov, Loren R. Graham, *Lysenko's Ghost: Epigenetics and Russia* (Cambridge, MA: Harvard University Press, 2016); Simon Ings, *Stalin and the Scientists: A History of Triumph and Tragedy, 1905–1953* (New York: Atlantic Monthly Press, 2016).

p. 157 "Deutsche Physik ... Jewish physics ... Philipp Lenard and Johannes Stark." Richard Rhodes, *The Making of the Atomic Bomb* (New York: Simon & Schuster, 1986), on the 1933 expulsions, the campaign against relativity, and the exiled physics that armed the Allies; and, for the movement itself and the parts played by Lenard and Stark, Alan D. Beyerchen, *Scientists under Hitler: Politics and the Physics Community in the Third Reich* (New Haven: Yale University Press, 1977).

p. 158 "the men who ran the tobacco companies ... to manufacture the look of a live scientific quarrel." Naomi Oreskes and Erik M. Conway, *Merchants of Doubt* (New York: Bloomsbury Press, 2010).

p. 160 "Barry Marshall … the bacterium … Barry Marshall stirred a culture of the bacterium into a cloudy broth and drank it down … ulcers." On the discovery that *Helicobacter pylori* causes peptic ulcers and Marshall's 1984 self-experiment, Barry J. Marshall and J. Robin Warren, "Unidentified Curved Bacilli in the Stomach of Patients with Gastritis and Peptic Ulceration," *The Lancet* 323, no. 8390 (1984): 1311–1315; the two shared the 2005 Nobel Prize in Physiology or Medicine.

Chapter Fifteen · The War Against Reality

p. 163 "The Four Pests campaign … the sparrow … the Great Leap … the famine." On the Great Leap Forward, the Four Pests campaign, and the famine of 1958–1962, Frank Dikötter, *Mao's Great Famine: The History of China's Most Devastating Catastrophe, 1958–1962* (New York: Walker & Company, 2010); Yang Jisheng, *Tombstone: The Great Chinese Famine, 1958–1962* (New York: Farrar, Straus and Giroux, 2012).

p. 163 "On the twelfth of February the campaign became national policy." Central Committee of the Chinese Communist Party and State Council, "Directive on Eliminating the Four Pests and Promoting Hygiene," 12 February 1958; the text ran on the front page of *People's Daily* the following day, 13 February 1958.

p. 164 "The writer Guo Moruo, one of the most celebrated men of letters in the country" Guo Moruo, "Zhou maque" 咒麻雀 ["Curse the Sparrows"], *Beijing Evening News*, 21 April 1958. The poem circulates widely as an anonymous anti-sparrow verse; the authorship and the date are established.

p. 164 "They drummed for two days without stopping … the Polish staff cleared the embassy grounds of dead birds with shovels." Zhiwu Frank, Wang, Wang, Wang, and You, "Campaigning for Extinction," National Bureau of Economic Research Working Paper 34087 (August 2025), n. 6, which sources the episode to *Gazeta Wyborcza*. Free PDF at DOI 10.3386/w34087.

p. 164 "The ornithologist Zheng Zuoxin knew what was in a sparrow's stomach, because he had opened sparrows and looked, and the answer was insects." Zheng published in English as Cheng Tien-Hsi; see Cheng Tien-Hsi, "Insect Control in Mainland China," *Science* 140, no. 3564 (19 April 1963): 269–277, for the stomach contents and the seasonal shift from grain to insects.

p. 164 "In August of 1956, at the national congress of the Chinese Zoological Society in Qingdao ... the cell biologist Zhu Xi rose first in the discussion of the sparrow." No contemporaneous transcript of Zhu Xi's remarks survives in the open record; the account here follows the reconstructions of the historian Xiong Weimin, drawn from the meeting papers and from participants' recollections, which agree on the occasion, on Zhu speaking first, and on the Prussian example he raised. The gathering was the second national congress of the Chinese Zoological Society, held at Qingdao, 23–28 August 1956. On the Prussian campaign itself, Frederick II, *Renovirtes und geschärfftes Edict wegen Ausrottung Der Sperlinge und Krähen*, Berlin, 22 June 1744, of which digitized copies survive in German university collections; the insect outbreaks that followed the killing, and the restocking that ended the policy, are standard in the German literature on the eighteenth-century *Spatzenkriege*.

p. 165 "a reported shipment of two hundred and fifty thousand sparrows imported from the Soviet Union." Frank et al., "Campaigning for Extinction," n. 8, which reports the figure and concedes that no formal validation of the purchase has been found. The hedge in the text is deliberate.

p. 165 "something close to a fifth of the collapse in crop yields and killed on its own about two million people." Frank et al., "Campaigning for Extinction," §5.5, which isolates the sparrow campaign from the other causes of the famine and puts its share of lost agricultural output at 19.64 percent and the resulting deaths at 1,954,169.

p. 167 "Lushan ... Peng Dehuai ... Mao." On the Lushan Conference of 1959 and Peng Dehuai's letter, Dikötter, *Mao's Great Famine*; Jung Chang and Jon Halliday, *Mao: The Unknown Story* (New York: Alfred A. Knopf, 2005).

p. 170 "Nearly every society in the human record ... was static." David Deutsch, *The Beginning of Infinity: Explanations That Transform the World* (New York: Viking, 2011), ch. 15, "The Evolution of Culture."

p. 171 "the photograph ... Yezhov ... airbrush ... re-touchers." On the removal of Nikolai Yezhov and others from Soviet photographs, David King, *The Commissar Vanishes: The Falsification of Photographs and Art in Stalin's Russia* (New York: Metropolitan Books, 1997).

p. 172 "It instructed them, in the calm bureaucratic prose of a routine update, to take a knife or a razor blade to their own encyclopedia."

The circular from the State Scientific Publishing House named pages 21, 22, 23 and 24 of volume five and the portrait bound between 22 and 23, to be cut out with scissors or a razor leaving a margin for the enclosed replacement pages, which expanded the adjacent articles on the Bering Sea and George Berkeley. The letter is reproduced in Masha Gessen, *The Future Is History: How Totalitarianism Reclaimed Russia* (New York: Riverhead Books, 2017); see also Joshua Rubenstein, *The Last Days of Stalin* (New Haven: Yale University Press, 2016), 226–227. Volume five had gone to press in September 1950; the circular reached subscribers in the first weeks of 1954.

p. 173 "The very concept of objective truth, he wrote, is fading out of the world." George Orwell, "Looking Back on the Spanish War" (written 1942; published 1943), which carries both this sentence and the prediction that a Russian army that never entered Spain would be written into the histories and learned as fact by schoolchildren generations later.

p. 173 "that two and two make five … a real Soviet poster of the First Five-Year Plan." Yakov Guminer's 1931 poster, "2 + 2 = 5: Arithmetic of a Counter-Plan Plus the Enthusiasm of the Workers." The likeliest route by which the slogan reached Orwell is Eugene Lyons, *Assignment in Utopia* (New York: Harcourt, Brace, 1937), whose chapter "Two Plus Two Equals Five" describes the formula as a Soviet commonplace.

Chapter Sixteen · Anyone Home

p. 175 "at sea in a dense fog … the most important of her life … there was no one yet home to mean it." Helen Keller, *The Story of My Life* (New York: Doubleday, Page & Co., 1903), ch. IV, which carries all three: the dense-fog image in her own words, the arrival of Anne Sullivan in March 1887, and the moment at the pump. The book is in the public domain; the free Project Gutenberg or Internet Archive text is the preferred citation.

p. 175 "Blake Lemoine … LaMDA." On the Google engineer Blake Lemoine's 2022 claim that the LaMDA language model was sentient, Nitasha Tiku, "The Google Engineer Who Thinks the Company's AI Has Come to Life," *Washington Post*, 11 June 2022.

p. 176 "In 1966 … MIT … Joseph Weizenbaum's secretary." Joseph Weizenbaum, *Computer Power and Human Reason: From Judgment to Calculation* (San Francisco: W. H. Freeman, 1976); and "ELIZA—A

Computer Program for the Study of Natural Language Communication between Man and Machine," *Communications of the ACM* 9 (1966): 36–45.

p. 176 "the philosopher Thomas Nagel … Nagel … the bat." Thomas Nagel, "What Is It Like to Be a Bat?" *The Philosophical Review* 83, no. 4 (1974): 435–450.

p. 178 "the zombie … the hard problem." David J. Chalmers, *The Conscious Mind: In Search of a Fundamental Theory* (New York: Oxford University Press, 1996); "Facing Up to the Problem of Consciousness," *Journal of Consciousness Studies* 2, no. 3 (1995): 200–219.

p. 179 "Dennett … The deflationary answer." Daniel C. Dennett, *Consciousness Explained* (Boston: Little, Brown, 1991).

p. 181 "an entire chapter of The Portrait of a Lady is a woman sitting motionless by a dying fire." Henry James, *The Portrait of a Lady* (1881), ch. 42, and his preface to the New York Edition (1908).

p. 181 "Seoul in March of 2016 … AlphaGo … Lee Sedol … its thirty-seventh move." David Silver et al., "Mastering the Game of Go with Deep Neural Networks and Tree Search," Nature 529 (2016): 484–489; and Greg Kohs, dir., *AlphaGo* (2017).

p. 181 "move seventy-eight … one in ten thousand … the hand of God." Cade Metz, *Genius Makers: The Mavericks Who Brought A.I. to Google, Facebook, and the World* (New York: Dutton, 2021), for Demis Hassabis's account of AlphaGo's own one-in-ten-thousand estimate of the move; and Greg Kohs, dir., *AlphaGo* (2017), which follows the fourth game in full.

p. 182 "It is a fighter for ends" and "a fighter for ends." William James, *The Principles of Psychology*, 2 vols. (New York: Henry Holt, 1890), vol. 1, chs. 1 and 5; the stream of consciousness, ch. 9.

p. 185 "I ought to be thy Adam, but I am rather the fallen angel." Mary Shelley, *Frankenstein; or, The Modern Prometheus* (London, 1818), vol. 2, ch. 7 (ch. 15 in the 1831 edition).

Chapter Seventeen · What the Lion Cannot Do

p. 189 "Gombe … Jane Goodall … until every male of the breakaway community had been hunted down and killed." Jane Goodall, *Through a Window: My Thirty Years with the Chimpanzees of Gombe*

(Boston: Houghton Mifflin, 1990); and *The Chimpanzees of Gombe: Patterns of Behavior* (Cambridge, MA: Belknap Press, 1986).

p. 190 "more than a thousand species of mammals ... near two percent of all deaths." José María Gómez, Miguel Verdú, Adela González-Megías, and Marcos Méndez, "The Phylogenetic Roots of Human Lethal Violence," Nature 538 (2016): 233–237; on the deep animal roots of moral behavior, Frans de Waal, *Primates and Philosophers: How Morality Evolved* (Princeton: Princeton University Press, 2006).

p. 191 "the fallibilist philosophy of Karl Popper, whom he had read and absorbed." Karl R. Popper, *The Logic of Scientific Discovery* (London: Hutchinson, 1959) and *Conjectures and Refutations* (London: Routledge, 1963); on the closed and the open society, *The Open Society and Its Enemies* (London: Routledge, 1945).

p. 194 "the British troopship HMS Birkenhead ... every woman and every child lived." A. C. Addison and W. H. Matthews, *A Deathless Story: The Birkenhead and Its Heroes* (London: Hutchinson, 1906); Rudyard Kipling, "Soldier an' Sailor Too," in *The Seven Seas* (1896).

p. 195 "An orca called Tahlequah ... seventeen days." On J35, known as Tahlequah, a southern resident killer whale who carried her dead calf for at least seventeen days and roughly a thousand miles in the summer of 2018, see the contemporaneous reports of the Center for Whale Research and Lynda V. Mapes's coverage in the *Seattle Times*.

p. 196 "Argos ... Boatswain ... Byron ... Kipling." Homer, *Odyssey*, bk. 17 (the dog Argos); Lord Byron, "Epitaph to a Dog" (1808), composed for his Newfoundland, Boatswain—the accompanying prose eulogy is by John Cam Hobhouse; Rudyard Kipling, "The Power of the Dog" (1909).

Chapter Eighteen · The Last Full Measure

p. 200 "Athens itself ... the little neutral island of Melos ... Thucydides, who recorded it." Thucydides, *History of the Peloponnesian War*, book 5, the Melian Dialogue (5.84–116); the sentence is at 5.89. Quoted here in the Richard Crawley translation of 1874, which is out of copyright and freely available; for a reader who wants maps and apparatus, Robert B. Strassler, ed., *The Landmark Thucydides* (New York: Free Press, 1996).

p. 202 "Lincoln ... he sat and wrote himself a private note resolving to spend the months he had left saving the union anyway." On Lincoln's "blind memorandum" of 23 August 1864, pledging to cooperate with his successor if he lost the election, Doris Kearns Goodwin, *Team of Rivals: The Political Genius of Abraham Lincoln* (New York: Simon & Schuster, 2005).

p. 202 "Washington ... Newburgh." On George Washington's address to his officers at Newburgh (15 March 1783), which quieted the threatened mutiny, Ron Chernow, *Washington: A Life* (New York: Penguin Press, 2010).

p. 203 "Told by the painter Benjamin West ... he would be the greatest man in the world." The exchange reaches us secondhand, through West's recollection as recorded by his biographer: John Galt, *The Life, Studies, and Works of Benjamin West* (London: T. Cadell and W. Davies, 1820), part 2. No contemporary record of George III's words survives, which is why the text says "is said to have answered"; Chernow, *Washington: A Life*, recounts the episode and its provenance.

p. 204 "Eisenhower ... Normandy ... wrote a second message, to be released only if the thing failed, taking all the blame." Eisenhower drafted the message on 5 June 1944, misdating it "July 5"; the original is held at the Dwight D. Eisenhower Presidential Library, Abilene, Kansas, and is reproduced in the National Archives' online exhibits. For the episode and its context, Jean Edward Smith, *Eisenhower in War and Peace* (New York: Random House, 2012).

Chapter Nineteen · The Radius

p. 209 "Valladolid ... Las Casas ... Sepúlveda." On the Valladolid debate of 1550–1551 between Bartolomé de las Casas and Juan Ginés de Sepúlveda, Lewis Hanke, *All Mankind Is One: A Study of the Disputation between Bartolomé de Las Casas and Juan Ginés de Sepúlveda* (DeKalb: Northern Illinois University Press, 1974).

p. 211 "In the spring of 1787, a dozen men met in a London printing shop and formed a society to abolish the slave trade." The Society for Effecting the Abolition of the Slave Trade was founded in May 1787. Thomas Clarkson, *The History of the Rise, Progress, and Accomplishment of the Abolition of the African Slave-Trade by the British Parliament*, 2 vols. (London: Longman, Hurst, Rees, and Orme, 1808), is both the fullest account of the committee's work and the source for Clarkson's years of evidence-gathering in the ports. In the public domain.

p. 211 "a plan of a real slave ship, the Brookes of Liverpool ... laid out in a space a coffin would have found tight." The print is "Stowage of the British Slave Ship Brookes under the Regulated Slave Trade Act of 1788." The Act fixed the licensed capacity at 454 and set the allocations the engraving renders: six feet by one foot four inches for an adult man, five feet ten by one foot four for a woman, five feet by one foot two for a boy. The engraving shows 482 figures against that limit, and the ship's own 1787 crossing had embarked 609, nineteen of whom died before Jamaica; the abolitionists' summary of the parliamentary evidence puts it as six hundred and nine by the confession of the slave-merchants in a subsequent voyage. *Abstract of the Evidence Delivered before a Select Committee of the House of Commons in the Years 1790 and 1791* (London, 1791). A Plymouth committee issued the first version of the broadside in 1788; the London revision followed. Clarkson, *History*, reproduces the print and records the period comparison of the allocated space to a coffin, so the phrase renders a contemporary judgment rather than a modern one.

p. 212 "No one saw it, Clarkson wrote, without being impressed; it spoke in a language at once intelligible and irresistible." Clarkson, *History*, vol. 2.

p. 212 "am I not a man and a brother ... pressed into brooches and set into hairpins and the lids of snuffboxes." On Josiah Wedgwood's cameo of the kneeling man in chains and its circulation as wearable political sentiment, Clarkson, *History*, vol. 2.

p. 212 "a Swiss businessman named Henri Dunant ... Solferino ... the Red Cross ... the first Geneva Convention." On Henri Dunant, the battle of Solferino (1859), and the founding of the Red Cross, Henri Dunant, *A Memory of Solferino* (1862; Geneva: International Committee of the Red Cross, 1986).

p. 214 "The expanding circle." Peter Singer, *The Expanding Circle: Ethics, Evolution, and Moral Progress* (New York: Farrar, Straus and Giroux, 1981); the image originates with W. E. H. Lecky, *History of European Morals from Augustus to Charlemagne* (1869).

Chapter Twenty · Backwards

p. 215 "Voyager ... Carl Sagan ... a mote of dust suspended in a sunbeam." On the "Pale Blue Dot" photograph (Voyager 1, 14 February

1990) and Sagan's reflection, Carl Sagan, *Pale Blue Dot: A Vision of the Human Future in Space* (New York: Random House, 1994).

p. 218 "a golden phonograph record ... Bach ... Chuck Berry ... fifty-five ... greetings ... Ann Druyan ... brainwaves ... the children of planet Earth." On the Voyager Golden Record—its music, its greetings in fifty-five languages, Ann Druyan's recorded brain waves, and the greeting spoken by Sagan's son Nick—Carl Sagan, F. D. Drake, Ann Druyan, Timothy Ferris, Jon Lomberg, and Linda Salzman Sagan, *Murmurs of Earth: The Voyager Interstellar Record* (New York: Random House, 1978).

BIBLIOGRAPHY

Abbey, Edward. *The Journey Home: Some Words in Defense of the American West.* New York: Dutton, 1977.

Abstract of the Evidence Delivered before a Select Committee of the House of Commons in the Years 1790 and 1791. London, 1791.

Addison, A. C., and W. H. Matthews. *A Deathless Story: The Birkenhead and Its Heroes.* London: Hutchinson, 1906.

Aeschylus. *Prometheus Bound.*

Aiello, Leslie C., and Peter Wheeler. "The Expensive-Tissue Hypothesis: The Brain and the Digestive System in Human and Primate Evolution." *Current Anthropology* 36, no. 2 (1995): 199–221.

Ainsworth, Mary D. Salter, Mary C. Blehar, Everett Waters, and Sally Wall. *Patterns of Attachment: A Psychological Study of the Strange Situation.* Hillsdale, NJ: Lawrence Erlbaum, 1978.

Arendt, Hannah. *Eichmann in Jerusalem: A Report on the Banality of Evil.* New York: Viking, 1963.

Aron, Arthur, Edward Melinat, Elaine N. Aron, Robert Darrin Vallone, and Renee J. Bator. "The Experimental Generation of Interpersonal Closeness: A Procedure and Some Preliminary Findings." *Personality and Social Psychology Bulletin* 23, no. 4 (1997): 363–377.

Augustine. *City of God.*

Bartlett, Frederic C. *Remembering: A Study in Experimental and Social Psychology.* Cambridge: Cambridge University Press, 1932.

Bastiat, Frédéric. "That Which Is Seen, and That Which Is Not Seen." 1850.

Beattie, Owen, and John Geiger. *Frozen in Time: The Fate of the Franklin Expedition.* Rev. ed. Vancouver: Greystone Books, 2004.

Benatar, David. *Better Never to Have Been: The Harm of Coming into Existence.* Oxford: Oxford University Press, 2006.

Berko, Jean. "The Child's Learning of English Morphology." *Word* 14, nos. 2–3 (1958): 150–177.

Beyerchen, Alan D. *Scientists under Hitler: Politics and the Physics Community in the Third Reich.* New Haven: Yale University Press, 1977.

Bilefsky, Dan. "Inky the Octopus Escapes From a New Zealand Aquarium." *New York Times*, April 14, 2016.

Bird, Kai, and Martin J. Sherwin. *American Prometheus: The Triumph and Tragedy of J. Robert Oppenheimer.* New York: Alfred A. Knopf, 2005.

Botkin, Daniel B. *Discordant Harmonies: A New Ecology for the Twenty-First Century.* New York: Oxford University Press, 1990.

Bowlby, John. *Attachment.* Vol. 1 of *Attachment and Loss.* New York: Basic Books, 1969.

Brickman, Philip, Dan Coates, and Ronnie Janoff-Bulman. "Lottery Winners and Accident Victims: Is Happiness Relative?" *Journal of Personality and Social Psychology* 36, no. 8 (1978): 917–927.

Bronowski, Jacob. *The Ascent of Man.* Boston: Little, Brown, 1973.

Buckley, Cara. "Earth Now Has 8 Billion Humans. This Man Wishes There Were None." *New York Times*, November 23, 2022.

Byron, Lord (George Gordon). "Epitaph to a Dog." 1808.

Cartailhac, Émile. "Les cavernes ornées de dessins. La grotte d'Altamira, Espagne. Mea culpa d'un sceptique." *L'Anthropologie* 13 (1902): 348ff.

Chalmers, David J. "Facing Up to the Problem of Consciousness." *Journal of Consciousness Studies* 2, no. 3 (1995): 200–219.

Chalmers, David J. *The Conscious Mind: In Search of a Fundamental Theory.* New York: Oxford University Press, 1996.

Chang, Jung, and Jon Halliday. *Mao: The Unknown Story.* New York: Alfred A. Knopf, 2005.

Cheng Tien-Hsi. "Insect Control in Mainland China." *Science* 140, no. 3564 (April 19, 1963): 269–277.

Chernow, Ron. *Washington: A Life.* New York: Penguin Press, 2010.

Chouinard, Michelle M. "Children's Questions: A Mechanism for Cognitive Development." *Monographs of the Society for Research in Child Development* 72, no. 1 (2007).

Clarkson, Thomas. *The History of the Rise, Progress, and Accomplishment of the Abolition of the African Slave-Trade by the British Parliament.* 2 vols. London: Longman, Hurst, Rees, and Orme, 1808.

Conard, Nicholas J., Maria Malina, and Susanne C. Münzel. "New Flutes Document the Earliest Musical Tradition in Southwestern Germany." *Nature* 460 (2009): 737–740.

Davis, Devra. *When Smoke Ran Like Water: Tales of Environmental Deception and the Battle against Pollution.* New York: Basic Books, 2002.

Dennett, Daniel C. *Consciousness Explained.* Boston: Little, Brown, 1991.

Deutsch, David. *The Beginning of Infinity: Explanations That Transform the World.* New York: Viking, 2011.

de Waal, Frans. *Primates and Philosophers: How Morality Evolved.* Princeton: Princeton University Press, 2006.

Dick, Philip K. *Do Androids Dream of Electric Sheep?* New York: Doubleday, 1968.

Dikötter, Frank. *Mao's Great Famine: The History of China's Most Devastating Catastrophe, 1958–1962.* New York: Walker & Company, 2010.

Dirac, P. A. M. "The Evolution of the Physicist's Picture of Nature." *Scientific American* 208, no. 5 (1963): 45–53.

Dissanayake, Ellen. *Homo Aestheticus: Where Art Comes from and Why.* New York: Free Press, 1992.

Dunant, Henri. *A Memory of Solferino.* 1862. Geneva: International Committee of the Red Cross, 1986.

Dunsworth, Holly M., Anna G. Warrener, Terrence Deacon, Peter T. Ellison, and Herman Pontzer. "Metabolic Hypothesis for Human Altriciality." *Proceedings of the National Academy of Sciences* 109, no. 38 (2012): 15212–15216.

Dutton, Denis. *The Art Instinct: Beauty, Pleasure, and Human Evolution.* New York: Bloomsbury Press, 2009.

Earl, John, and W. J. McCleary. "Mystery of the Poisoned Expedition." *Nature* 368 (1994): 683–684.

Ehrlich, Paul R. *The Population Bomb.* New York: Ballantine, 1968.

Epstein, Alex. *Fossil Future: Why Global Human Flourishing Requires More Oil, Coal, and Natural Gas—Not Less.* New York: Portfolio, 2022.

Epstein, Alex. *The Moral Case for Fossil Fuels.* New York: Portfolio, 2014.

Farman, J. C., B. G. Gardiner, and J. D. Shanklin. "Large Losses of Total Ozone in Antarctica Reveal Seasonal ClOx/NOx Interaction." *Nature* 315 (1985): 207–210.

Farmelo, Graham. *The Strangest Man: The Hidden Life of Paul Dirac.* New York: Basic Books, 2009.

Festinger, Leon. *A Theory of Cognitive Dissonance.* Stanford: Stanford University Press, 1957.

Festinger, Leon, Henry W. Riecken, and Stanley Schachter. *When Prophecy Fails.* Minneapolis: University of Minnesota Press, 1956.

Feynman, Richard P. "Personal Observations on the Reliability of the Shuttle." Appendix F to the *Report of the Presidential Commission on the Space Shuttle Challenger Accident.* Washington, DC, 1986.

Feynman, Richard P. *The Character of Physical Law.* Cambridge, MA: MIT Press, 1965.

Feynman, Richard P. *What Do You Care What Other People Think?* New York: W. W. Norton, 1988.

Fishman, Charles. *One Giant Leap: The Impossible Mission That Flew Us to the Moon.* New York: Simon & Schuster, 2019.

Frank, Richard B. *Downfall: The End of the Imperial Japanese Empire.* New York: Random House, 1999.

Gardner, R. Allen, and Beatrice T. Gardner. "Teaching Sign Language to a Chimpanzee." *Science* 165 (1969): 664–672.

Gessen, Masha. *The Future Is History: How Totalitarianism Reclaimed Russia.* New York: Riverhead Books, 2017.

Giangreco, D. M. "Half a Million Purple Hearts." *American Heritage* 51, no. 8 (December 2000).

Gilbert, Daniel. *Stumbling on Happiness.* New York: Alfred A. Knopf, 2006.

Gilovich, Thomas, and Victoria Husted Medvec. "The Experience of Regret: What, When, and Why." *Psychological Review* 102, no. 2 (1995): 379–395.

Gimpel, Jean. *The Cathedral Builders.* Translated by Teresa Waugh. New York: Grove Press, 1983.

Gómez, José María, Miguel Verdú, Adela González-Megías, and Marcos Méndez. "The Phylogenetic Roots of Human Lethal Violence." *Nature* 538 (2016): 233–237.

Goodall, Jane. *Through a Window: My Thirty Years with the Chimpanzees of Gombe.* Boston: Houghton Mifflin, 1990.

Goodwin, Doris Kearns. *Team of Rivals: The Political Genius of Abraham Lincoln.* New York: Simon & Schuster, 2005.

Gopnik, Alison, and David M. Sobel. "Detecting Blickets: How Young Children Use Information about Novel Causal Powers in Categorization and Induction." *Child Development* 71, no. 5 (2000): 1205–1222.

Gopnik, Alison. "Childhood as a Solution to Explore–Exploit Tensions." *Philosophical Transactions of the Royal Society B* 375, no. 1803 (2020): 20190502.

Gopnik, Alison, Clark Glymour, David M. Sobel, Laura E. Schulz, Tamar Kushnir, and David Danks. "A Theory of Causal Learning in Children: Causal Maps and Bayes Nets." *Psychological Review* 111, no. 1 (2004): 3–32.

Gopnik, Alison. *The Gardener and the Carpenter: What the New Science of Child Development Tells Us about the Relationship between Parents and Children.* New York: Farrar, Straus and Giroux, 2016.

Gopnik, Alison. *The Philosophical Baby: What Children's Minds Tell Us about Truth, Love, and the Meaning of Life.* New York: Farrar, Straus and Giroux, 2009.

Gottman, John, and Joan DeClaire. *The Relationship Cure.* New York: Crown, 2001.

Gottman, John M., and Nan Silver. *The Seven Principles for Making Marriage Work.* New York: Crown, 1999.

Gottman, John M., and Robert W. Levenson. "The Timing of Divorce: Predicting When a Couple Will Divorce over a 14-Year Period." *Journal of Marriage and Family* 62, no. 3 (2000): 737–745.

Graham, Loren R. *Lysenko's Ghost: Epigenetics and Russia.* Cambridge, MA: Harvard University Press, 2016.

Guo Moruo. "Zhou maque" 咒麻雀 ["Curse the Sparrows"]. *Beijing Evening News*, April 21, 1958.

Hanke, Lewis. *All Mankind Is One: A Study of the Disputation between Bartolomé de Las Casas and Juan Ginés de Sepúlveda.* DeKalb: Northern Illinois University Press, 1974.

Harris, Paul L. *The Work of the Imagination.* Oxford: Blackwell, 2000.

Hasegawa, Tsuyoshi. Racing the Enemy: Stalin, Truman, and the Surrender of Japan. Cambridge, MA: Harvard University Press, 2005.

Hassabis, Demis, Dharshan Kumaran, Seralynne D. Vann, and Eleanor A. Maguire. "Patients with Hippocampal Amnesia Cannot Imagine New Experiences." *Proceedings of the National Academy of Sciences* 104, no. 5 (2007): 1726–1731.

Hauser, Marc D., Noam Chomsky, and W. Tecumseh Fitch. "The Faculty of Language: What Is It, Who Has It, and How Did It Evolve?" *Science* 298 (2002): 1569–1579.

Hawkes, Kristen, James F. O'Connell, and Nicholas G. Blurton Jones. "Hadza Women's Time Allocation, Offspring Provisioning, and the Evolution of Long Postmenopausal Life Spans." *Current Anthropology* 38, no. 4 (1997): 551–577.

Hayek, Friedrich A. "The Use of Knowledge in Society." *American Economic Review* 35, no. 4 (1945): 519–530.

Henderson, D. A. *Smallpox: The Death of a Disease.* Amherst, NY: Prometheus Books, 2009.

Henshilwood, Christopher S., et al. "Emergence of Modern Human Behaviour: Middle Stone Age Engravings from South Africa." *Science* 295, no. 5558 (2002): 1278–1280.

Hesiod. *Theogony* and *Works and Days.*

Hesser, Leon. *The Man Who Fed the World.* Dallas: Durban House, 2006.

Hirschman, Albert O. *The Passions and the Interests: Political Arguments for Capitalism before Its Triumph.* Princeton: Princeton University Press, 1977.

Hoeksema, Robert J. *Designed for Dry Feet: Flood Protection and Land Reclamation in the Netherlands.* Reston, VA: ASCE Press, 2006.

Holt-Lunstad, Julianne, Timothy B. Smith, and J. Bradley Layton. "Social Relationships and Mortality Risk: A Meta-analytic Review." *PLoS Medicine* 7, no. 7 (2010): e1000316.

Homer. *The Odyssey.*

Horner, Victoria, and Andrew Whiten. "Causal Knowledge and Imitation/Emulation Switching in Chimpanzees (*Pan troglodytes*) and Children (*Homo sapiens*)." *Animal Cognition* 8, no. 3 (2005): 164–181.

Hrdy, Sarah Blaffer. *Mothers and Others: The Evolutionary Origins of Mutual Understanding.* Cambridge, MA: Belknap Press of Harvard University Press, 2009.

Hubel, David H., and Torsten N. Wiesel. "The Period of Susceptibility to the Physiological Effects of Unilateral Eye Closure in Kittens." *Journal of Physiology* 206, no. 2 (1970): 419–436.

Hunt, Gavin R. "Manufacture and Use of Hook-Tools by New Caledonian Crows." *Nature* 379 (1996): 249–251.

Hülsmann, Jörg Guido. *Mises: The Last Knight of Liberalism.* Auburn, AL: Ludwig von Mises Institute, 2007.

Ings, Simon. *Stalin and the Scientists: A History of Triumph and Tragedy, 1905–1953.* New York: Atlantic Monthly Press, 2016.

James, Henry. *The Portrait of a Lady.* 1881.

James, William. *The Principles of Psychology.* 2 vols. New York: Henry Holt, 1890.

Jones, Darryl N., René W. R. J. Dekker, and Cees S. Roselaar. *The Megapodes: Megapodiidae.* Oxford: Oxford University Press, 1995.

Joyce, Chip. *The Fortress: Recovering the Bill of Rights as a Defense Against Tyranny.* Cornelius, NC: Cardo Press, 2026.

Junkin, Tim. *Bloodsworth: The True Story of the First Death Row Inmate Exonerated by DNA.* Chapel Hill, NC: Algonquin Books, 2004.

Kaplan, Robert. *The Nothing That Is: A Natural History of Zero.* Oxford: Oxford University Press, 2000.

Keats, John. "Ode on a Grecian Urn." 1819.

Keller, Helen. *The Story of My Life.* New York: Doubleday, Page & Co., 1903.

Kennedy, John F. "Address at Rice University on the Nation's Space Effort." Houston, TX, September 12, 1962.

Killingsworth, Matthew A., and Daniel T. Gilbert. "A Wandering Mind Is an Unhappy Mind." *Science* 330, no. 6006 (2010): 932.

Kind, Claus-Joachim, Nicole Ebinger-Rist, Sibylle Wolf, Thomas Beutelspacher, and Kurt Wehrberger. "The Smile of the Lion Man: Recent Excavations in Stadel Cave (Baden-Württemberg, South-Western Germany) and the Restoration of the Famous Upper Palaeolithic Figurine." *Quartär* 61 (2014).

King, David. *The Commissar Vanishes: The Falsification of Photographs and Art in Stalin's Russia.* New York: Metropolitan Books, 1997.

Kipling, Rudyard. "The Power of the Dog." 1909.

Klump, Barbara C., et al. "Hook Tool Manufacture in New Caledonian Crows: Behavioural Variation and the Influence of Raw Materials." *BMC Biology* 13 (2015): 97.

Kohs, Greg, dir. *AlphaGo.* 2017.

Lecky, W. E. H. *History of European Morals from Augustus to Charlemagne.* 1869.

Loftus, Elizabeth F. "Planting Misinformation in the Human Mind: A 30-Year Investigation of the Malleability of Memory." *Learning & Memory* 12, no. 4 (2005): 361–366.

Lyons, Derek E., Andrew G. Young, and Frank C. Keil. "The Hidden Structure of Overimitation." *Proceedings of the National Academy of Sciences* 104, no. 50 (2007): 19751–19756.

Madariaga de la Campa, Benito. *Sanz de Sautuola and the Discovery of the Caves of Altamira: Some Observations on the Paintings.* Translated by Shirley Clarke. Santander: Fundación Marcelino Botín, 2001.

Madison, James, Alexander Hamilton, and John Jay. *The Federalist Papers.* 1787–1788.

Mann, Charles C. *The Wizard and the Prophet: Two Remarkable Scientists and Their Dueling Visions to Shape Tomorrow's World.* New York: Alfred A. Knopf, 2018.

Marcus, Gary F., Steven Pinker, Michael Ullman, Michelle Hollander, T. John Rosen, and Fei Xu. "Overregularization in Language Acquisition." *Monographs of the Society for Research in Child Development* 57, no. 4 (1992).

Marshall, Barry J., and J. Robin Warren. "Unidentified Curved Bacilli in the Stomach of Patients with Gastritis and Peptic Ulceration." *The Lancet* 323, no. 8390 (1984): 1311–1315.

May, Todd. *Should We Go Extinct? A Philosophical Dilemma for Our Unbearable Times.* New York: Crown, 2024.

Mercader, Julio, et al. "4,300-Year-Old Chimpanzee Sites and the Origins of Percussive Stone Technology." *Proceedings of the National Academy of Sciences* 104, no. 9 (2007): 3043–3048.

Metz, Cade. *Genius Makers: The Mavericks Who Brought A.I. to Google, Facebook, and the World.* New York: Dutton, 2021.

Miller, Geoffrey. *The Mating Mind: How Sexual Choice Shaped the Evolution of Human Nature.* New York: Doubleday, 2000.

Milton, John. *Paradise Lost.* 1667.

Mintz, Morton. "'Heroine' of FDA Keeps Bad Drug Off Market." *Washington Post,* July 15, 1962.

Mises, Ludwig von. "Economic Calculation in the Socialist Commonwealth." 1920. Translated by S. Adler. Auburn, AL: Ludwig von Mises Institute, 1990.

Mises, Ludwig von. *Human Action: A Treatise on Economics.* New Haven: Yale University Press, 1949.

Mises, Ludwig von. *Liberalism: The Classical Tradition.* Translated by Ralph Raico. 1927. Indianapolis: Liberty Fund, 2005.

Molina, Mario J., and F. S. Rowland. "Stratospheric Sink for Chlorofluoromethanes: Chlorine Atom-Catalysed Destruction of Ozone." *Nature* 249 (1974): 810–812.

Moorehead, Alan. *Cooper's Creek.* London: Hamish Hamilton, 1963.

Musk, Elon. "Making Humans a Multi-Planetary Species." *New Space* 5, no. 2 (2017): 46–61.

Nagel, Thomas. "What Is It Like to Be a Bat?" *The Philosophical Review* 83, no. 4 (1974): 435–450.

Nuland, Sherwin B. *The Doctors' Plague: Germs, Childbed Fever, and the Strange Story of Ignác Semmelweis.* New York: W. W. Norton, 2003.

Oakley, Kenneth P. *Man the Tool-Maker.* London: British Museum, 1949.

Oreskes, Naomi, and Erik M. Conway. *Merchants of Doubt.* New York: Bloomsbury Press, 2010.

Orians, Gordon H., and Judith H. Heerwagen. "Evolved Responses to Landscapes." In *The Adapted Mind,* edited by Jerome H. Barkow, Leda

Cosmides, and John Tooby, 555–579. New York: Oxford University Press, 1992.

Orwell, George. "Looking Back on the Spanish War." 1943.

Otis, James. *The Rights of the British Colonies Asserted and Proved*. Boston: Edes and Gill, 1764.

Pascal, Blaise. *Pensées*. Translated by A. J. Krailsheimer. Harmondsworth: Penguin, 1966.

Peikoff, Leonard. *Objectivism: The Philosophy of Ayn Rand*. New York: Dutton, 1991.

Pepperberg, Irene M. *Alex & Me*. New York: Collins, 2008.

Pepperberg, Irene M., and Jesse D. Gordon. "Number Comprehension by a Grey Parrot (*Psittacus erithacus*), Including a Zero-Like Concept." *Journal of Comparative Psychology* 119, no. 2 (2005): 197–209.

Pepperberg, Irene M. *The Alex Studies: Cognitive and Communicative Abilities of Grey Parrots*. Cambridge, MA: Harvard University Press, 1999.

Piaget, Jean. *The Child's Conception of Number*. Translated by C. Gattegno and F. M. Hodgson. London: Routledge & Kegan Paul, 1952.

Pinker, Steven. *How the Mind Works*. New York: W. W. Norton, 1997.

Pinker, Steven. *Words and Rules: The Ingredients of Language*. New York: Basic Books, 1999.

Pomeroy, Emma, et al. "New Neanderthal Remains Associated with the 'Flower Burial' at Shanidar Cave." *Antiquity* 94, no. 373 (2020): 11–26.

Popper, Karl R. *Conjectures and Refutations: The Growth of Scientific Knowledge*. London: Routledge, 1963.

Popper, Karl R. *The Logic of Scientific Discovery*. London: Hutchinson, 1959.

Popper, Karl R. *The Open Society and Its Enemies*. London: Routledge, 1945.

Portmann, Adolf. *A Zoologist Looks at Humankind*. Translated by Judith Schaefer. New York: Columbia University Press, 1990.

Raffaele, Paul. "Speaking Bonobo." *Smithsonian*, November 2006.

Rand, Ayn. *The Romantic Manifesto: A Philosophy of Literature*. Rev. ed. New York: Signet, 1975.

Rand, Ayn. "The Objectivist Ethics." In *The Virtue of Selfishness*. New York: New American Library, 1964.

Read, Leonard E. "I, Pencil." *The Freeman*, December 1958.

Rhodes, Richard. *The Making of the Atomic Bomb*. New York: Simon & Schuster, 1986.

Rilke, Rainer Maria. *Letters to a Young Poet*. Translated by M. D. Herter Norton. New York: W. W. Norton, 1934.

Rubenstein, Joshua. *The Last Days of Stalin*. New Haven: Yale University Press, 2016.

Sabin, Paul. *The Bet: Paul Ehrlich, Julian Simon, and Our Gamble over Earth's Future*. New Haven: Yale University Press, 2013.

Sagan, Carl, F. D. Drake, Ann Druyan, Timothy Ferris, Jon Lomberg, and Linda Salzman Sagan. *Murmurs of Earth: The Voyager Interstellar Record*. New York: Random House, 1978.

Sagan, Carl. *Pale Blue Dot: A Vision of the Human Future in Space*. New York: Random House, 1994.

Savage-Rumbaugh, E. Sue, et al. "Language Comprehension in Ape and Child." *Monographs of the Society for Research in Child Development* 58, nos. 3–4 (1993).

Schacter, Daniel L., and Donna Rose Addis. "The Cognitive Neuroscience of Constructive Memory: Remembering the Past and Imagining the Future." *Philosophical Transactions of the Royal Society B* 362, no. 1481 (2007): 773–786.

Semmelweis, Ignaz. *The Etiology, Concept, and Prophylaxis of Childbed Fever*. 1861. Translated by K. Codell Carter. Madison: University of Wisconsin Press, 1983.

Shelley, Mary. *Frankenstein; or, The Modern Prometheus*. 1818.

Shelley, Percy Bysshe. *Prometheus Unbound*. 1820.

Silver, David, et al. "Mastering the Game of Go with Deep Neural Networks and Tree Search." *Nature* 529 (2016): 484–489.

Simon, Julian L. *The Ultimate Resource*. Princeton: Princeton University Press, 1981.

Singer, Peter. *The Expanding Circle: Ethics, Evolution, and Moral Progress*. New York: Farrar, Straus and Giroux, 1981.

Smith, Jean Edward. *Eisenhower in War and Peace.* New York: Random House, 2012.

Solecki, Ralph S. *Shanidar: The First Flower People.* New York: Alfred A. Knopf, 1971.

Sommer, Jeffrey D. "The Shanidar IV 'Flower Burial': A Reevaluation of Neanderthal Burial Ritual." *Cambridge Archaeological Journal* 9, no. 1 (1999): 127–129.

Sowell, Thomas. *Applied Economics: Thinking beyond Stage One.* 2nd ed. New York: Basic Books, 2009.

Sowell, Thomas. *Basic Economics: A Common Sense Guide to the Economy.* 5th ed. New York: Basic Books, 2015.

Sowell, Thomas. *Knowledge and Decisions.* New York: Basic Books, 1980.

Spelke, Elizabeth S., and Katherine D. Kinzler. "Core Knowledge." *Developmental Science* 10, no. 1 (2007): 89–96.

Stupple, Aaron, with Logan Chipkin. *The Sovereign Child: How a Forgotten Philosophy Can Liberate Kids and Their Parents.* 2025.

Suddendorf, Thomas, and Michael C. Corballis. "The Evolution of Foresight: What Is Mental Time Travel, and Is It Unique to Humans?" *Behavioral and Brain Sciences* 30, no. 3 (2007): 299–351.

Tennie, Claudio, Josep Call, and Michael Tomasello. "Ratcheting Up the Ratchet: On the Evolution of Cumulative Culture." *Philosophical Transactions of the Royal Society B* 364, no. 1528 (2009): 2405–2415.

Terrace, Herbert S., Laura A. Petitto, R. J. Sanders, and Thomas G. Bever. "Can an Ape Create a Sentence?" *Science* 206 (1979): 891–902.

The Bhagavad Gita.

The Holy Bible (Genesis, Psalms, Romans).

Tiku, Nitasha. "The Google Engineer Who Thinks the Company's AI Has Come to Life." *Washington Post,* June 11, 2022.

Thucydides. *History of the Peloponnesian War.* Translated by Richard Crawley. 1874.

Tomasello, Michael. *The Cultural Origins of Human Cognition.* Cambridge, MA: Harvard University Press, 1999.

Tronick, Edward, Heidelise Als, Lauren Adamson, Susan Wise, and T. Berry Brazelton. "The Infant's Response to Entrapment between Contradictory Messages in Face-to-Face Interaction." *Journal of the American Academy of Child Psychiatry* 17, no. 1 (1978): 1–13.

U.S. Surgeon General. *Our Epidemic of Loneliness and Isolation.* Washington, DC: Office of the U.S. Surgeon General, 2023.

Volavková, Hana, ed. *I Never Saw Another Butterfly: Children's Drawings and Poems from Terezín Concentration Camp, 1942–1944.* Expanded 2nd ed. New York: Schocken Books, 1993.

Von Drehle, David. *Triangle: The Fire That Changed America.* New York: Atlantic Monthly Press, 2003.

Wallace, Charles, Jr., ed. *Susanna Wesley: The Complete Writings.* New York: Oxford University Press, 1997.

Washburn, Sherwood L. "Tools and Human Evolution." *Scientific American* 203, no. 3 (September 1960): 62–75.

Watson, John B. *Psychological Care of Infant and Child.* New York: W. W. Norton, 1928.

Weinberg, Steven. *Dreams of a Final Theory.* New York: Pantheon, 1992.

Weir, Alex A. S., Jackie Chappell, and Alex Kacelnik. "Shaping of Hooks in New Caledonian Crows." *Science* 297, no. 5583 (2002): 981.

Weisman, Alan. *The World Without Us.* New York: St. Martin's Press, 2007.

Weizenbaum, Joseph. *Computer Power and Human Reason: From Judgment to Calculation.* San Francisco: W. H. Freeman, 1976.

Weizenbaum, Joseph. "ELIZA—A Computer Program for the Study of Natural Language Communication between Man and Machine." *Communications of the ACM* 9, no. 1 (1966): 36–45.

Wesley, John. *The Journal of the Rev. John Wesley.* Edited by Nehemiah Curnock. London: Epworth Press, 1909–1916.

Wrangham, Richard. *Catching Fire: How Cooking Made Us Human.* New York: Basic Books, 2009.

Wynn, Karen. "Addition and Subtraction by Human Infants." *Nature* 358 (1992): 749–750.

Yang, Jisheng. *Tombstone: The Great Chinese Famine, 1958–1962.* New York: Farrar, Straus and Giroux, 2012.

FURTHER READING

The notes at the back of this book record where its claims come from. This list is a different thing. These are the books to read next—the ones that shaped the argument, or that open the door to a thread you may want to follow further. They are grouped, not ranked, though within each group the first title is the place to begin.

How Knowledge Grows

Deutsch, David. *The Beginning of Infinity.* 2011. — Deutsch argues that knowledge grows by a single process wherever it grows at all—conjecture tested by criticism—that good explanations are hard to vary, and that this process has no natural end. Start here if you start anywhere.

Popper, Karl. *Conjectures and Refutations.* 1963. — Where the engine was first named. Popper's case that we never prove a theory but only fail, so far, to refute it, and that this is the whole of the method, is the argument in its original and still-electric form.

Bronowski, Jacob. *The Ascent of Man.* 1973. — Proof that science can be told as a human story with no loss of rigor. The chapter "Knowledge or Certainty," which ends with Bronowski wading into the pond at Auschwitz where members of his family were killed, is the moral center of this whole tradition.

Reason, Values, and the Vision of Man

Rand, Ayn. *The Virtue of Selfishness.* 1964. — The ethics at full strength: that reason is a human being's basic means of survival, and that values can be discovered rather than decreed.

Peikoff, Leonard. *Objectivism: The Philosophy of Ayn Rand.* 1991. — The systematic statement Rand never wrote herself: the whole philosophy laid out in order, from perception to politics. He was my professor; the dedication is his.

Sowell, Thomas. *A Conflict of Visions.* 1987. — The fault line beneath two centuries of political argument: whether human nature is a fixed constraint to be managed, or something that can be corrected and improved. Read it beside this book's introduction; it names the divide the whole case turns on.

The Cooperation of Strangers

Mises, Ludwig von. *Liberalism.* 1927. — The short, humane one, and the place to begin with Mises. Its central discovery—that a free stranger trading with you is worth more than a conquered one, which is the real arithmetic of peace—is put with a clarity the larger works rarely pause for.

Mises, Ludwig von. *Human Action.* 1949. — The full Austrian system, forbidding in length and unmatched in reach. The 1920 argument that a planned economy is blind, not merely unfair, stands here in its mature form.

Hayek, Friedrich. *Individualism and Economic Order.* 1948. — Hayek's essays on how a market thinks, especially "The Use of Knowledge in Society" (1945)—fifteen pages that changed economics, and the place to meet firsthand the insight that a price is a signal carrying knowledge no single mind holds.

Sowell, Thomas. *Basic Economics.* — The plainest tour of the whole subject, without a graph or a jargon word in it: the book for anyone who was ever told economics was above them.

Sowell, Thomas. *Knowledge and Decisions.* 1980. — Sowell's own deepest work, carrying Hayek further: about where knowledge sits in a society, and who is permitted to act on it. Slower and more rewarding than the popular titles.

Bastiat, Frédéric. *Selected Essays on Political Economy.* — Bastiat at his sharpest, especially "What Is Seen and What Is Not Seen" (1850)—the original lesson in consequence, the visible benefit set against the cost no one photographs. An afternoon's reading that reorganizes how you see every law.

Epstein, Alex. *The Moral Case for Fossil Fuels.* 2014. — Here for its method more than its conclusions, sound as they are. Epstein's insistence on naming the standard—better for whom, and by what measure?—before the argument begins is the transferable tool, and a rare one.

The Human Story

Hrdy, Sarah Blaffer. *Mothers and Others.* 2009. — The science beneath "the second womb." Hrdy's case that humans are cooperative breeders—reared by many hands long before we were clever and made clever partly by it—is a landmark, and a good read.

Gopnik, Alison. *The Philosophical Baby.* 2009. — The child as scientist, told by the researcher who ran many of the experiments. If the nursery chapters held you, read this next; it also makes, quietly, the case behind the parenting chapter.

Wrangham, Richard. *Catching Fire.* 2009. — One of the best short science books here: the argument that cooking, and not merely meat, is what grew the human brain. It settles into the mind like something you always knew.

Dutton, Denis. *The Art Instinct.* 2009. — The serious and accessible case that the sense of beauty is old equipment, carried out of deep time and standard-issue in the species, rather than a late refinement of culture.

The Refusal to See

Arendt, Hannah. *Eichmann in Jerusalem.* 1963. — The report that gave us "the banality of evil": the finding that the administrator of a genocide was, above all, a man who would not think from anyone else's place. Argued over since the day it appeared, and still indispensable.

Bird, Kai, and Martin Sherwin. *American Prometheus.* 2005. — The definitive life of Oppenheimer, and the fullest account we have of what it does to a mind to build the unthinkable and then live inside the knowing. Long, and worth every page.

Dikötter, Frank. *Mao's Great Famine.* 2010. — The archive-based reckoning with what follows when a regime makes war on reality and forbids the correction: tens of millions dead of a lie about grain. Harrowing, and the ground truth beneath the war-on-reality chapter.

A Last One

Sagan, Carl. *Pale Blue Dot.* 1994. — If the last pages of this book found you, end where they end. Sagan's photograph, and the passage it drew from him, are the register this whole argument reaches toward: the smallness of our world set against the size of what looks out from it.

INDEX

www.ingramcontent.com/pod-product-compliance
Lightning Source LLC
LaVergne TN
LVHW100726061126
851857LV00011B/414
9798997402013